SIXTY YEARS OF BIOLOGY

JOHN TYLER BONNER

Sixty Years of Biology

Essays on Evolution and Development

PRINCETON UNIVERSITY PRESS

Published by Princeton University Press, 41 William Street,
Princeton, New Jersey 08540
In the United Kingdom: Princeton University Press,
Chichester, West Sussex

Chapter 5 first appeared as "From Darwin to DNA"
in the *Times Literary Supplement*. Reprinted with Permission.

Library of Congress Cataloging-in-Publication Data

Bonner, John Tyler.
Sixty years of biology : essays on evolution
and development / John Tyler Bonner.
p. cm.
Includes bibliographical references (p.) and index.
ISBN 0-691-02130-9 (alk. paper)
1. Evolution (Biology) 2. Natural selection.
3. Developmental biology. 4. Biology I. Title.
QH366.2.B666 1996
575—dc20 95-25638

This book has been composed in Palatino

Princeton University Press books are printed on
acid-free paper and meet the guidelines for permanence
and durability of the Committee on Production
Guidelines for Book Longevity of the
Council on Library Resources

Printed in the United States of America
by Princeton Academic Press

2 4 6 8 10 9 7 5 3 1

CONTENTS

PREFACE

I HAVE spent a good part of my scientific life trying to look at standard themes in a new, nonstandard light. One way I try to do this is by casting my net as far as possible and looking for comparisons at the extremes of the evolutionary scale, from molecules to animal behavior, and all that goes in between. Looking at the big picture is not a conventional approach, particularly in this age of intense specialization. And let me quickly add that I am not a philosopher but a biologist, and as the years advance I find myself less and less interested in the philosophical aspects of my large sweeps into the living world. Again this is not the standard approach where the philosophy of evolutionary biology has become a popular field of study. Clearly I do not enjoy running with the pack, and writing books is the best outlet for a scientific misfit like myself.

The contents of this book have been fermenting in my mind for a long time. Let me caution that there is no central theme to these essays—they simply represent different aspects of development and evolution that I have been trying to clarify in my own mind. Many of the themes will be well known to the reader, and some I have discussed in previous books. What I have tried to do here is to look at old problems in new ways: in some cases that involves genuine novelty, while in others the novelty will reside solely in stating the case with greater clarity. Sometimes describing the obvious in a fresh way can help one towards a deeper understanding.

The first chapter is on self-organization and natural selection. For me the subject goes back over thirty-five years, when I had the temerity to do an abridged edition of D'Arcy Thompson's *On Growth and Form*. That was my first serious encounter with the thought that there could be separate worlds of Darwinian selection and the physics of form. My view today is that they are of equal importance and the interesting thing is how they fit together.

Chapter 2 is a panoramic view of how competition has shaped the inner organization of living entities at all levels, spanning the gamut from primitive cells to human behavior. What is so interesting is the great sameness at each level. What novelty there is in this chapter comes not from the basic facts, but from the act of stringing them all together.

In the third chapter there is some heresy. The changes back and forth from genetically fixed living activities to more flexible ones has been a subject that has fascinated me for years: it has such direct application to problems of development and behavior. The way in which the necessary genetic accommodations for such shifts are made is not a subject commonly addressed. Here I suggest mechanisms that might please some and offend others.

In chapter 4 I return to an old theme that I have written about before, namely the division of labor. By the use of new examples at the various levels in the evolutionary scale, and in particular by looking at the incipient stages of dividing the labor in social amoebae and social insects, I have tried to deepen my own insight into why and how the labor is divided, and can only hope that I do so for the reader as well.

Finally, in the last chapter I take advantage of my numerous years as a biologist to look at the volcanic changes that have appeared in the subject since I was a beginner. This is a personal odyssey which does have its dangers—I once read a devastating review of a book which said that the author was in his "anecdotage." It is true the chapter is laced with anecdotes and reminiscences, but I am hoping that it goes beyond them and gives a convincing view of how our understanding of biology during most of this century has developed in such extraordinary ways. I feel as though I have been standing near the crater and have watched the volcano go through a succession of gargantuan eruptions.

IT SEEMS to me that to an unusual degree I am utterly dependent on my friends for comments on early versions of anything I write. With this book I blush to think of the rawness of the first drafts that they had to suffer. Living in one's own thought world can be fun and even exciting at times, but for me it is exceedingly dangerous, and therefore my gratitude to my helpers for saving me from myself is enormous. In particular, for their great labors I want to thank Henry Horn, Leo Buss, and Carlos Montinez del Rio. I have not agreed with all their criticisms (I wanted to leave something to stir the reader), but all made me think more deeply on many matters and for that I am truly grateful. I also wish to thank Ted Cox and Diane O'Brian for their help with the first chapter, and Jonathan Weiner for some significant help at the last stages. I am particularly indebted to Steven Rubin for his careful and most helpful scrutiny of the entire manuscript. Finally, I want to thank Emily Wilkinson for her continuing support of my writing efforts, and Alice Calaprice for again using her masterful skills in keeping my sentences from bulging in the wrong places. There are many other colleagues and friends who have contributed in other ways, such as in conversations and in answering questions—that list is very long and I thank them all.

August 1995
Margaree Harbour
Cape Breton, Nova Scotia

SIXTY YEARS OF BIOLOGY

CHAPTER 1

Self-Organization and Natural Selection

IT WAS D'Arcy Thompson who made us see clearly the importance of physical forces in the construction of living organisms. He showed that to a remarkable degree the form of animals and plants could be described in physical and mathematical terms—that nature subscribed to the sound principles of engineering. The shape of trees, the structure of the skeletons of mammals, the orientation of the trabeculae of bones, the shapes of horns, of gastropod shells, and many other living forms—all are built in such a way that they are optimal in their design for their particular function. As was true of so many of his generation, he was down on natural selection, except as a means of eliminating undesirable traits. He viewed all the forms of life to be the result of these inner physical forces, although he did draw a line. In his opening chapter he says, "Of how it is that the soul informs the body, physical science teaches me nothing," and later in the same paragraph, "nor do I ask of physics how goodness shines in one man's face, and evil betrays itself in another." Here I do not want to enter into such realms, either, but wish to confine myself to a much more straightforward problem. How much of living form can be explained by physical and chemical forces directly expressing themselves, and how much can be explained by their being reigned in by natural selection?

I will not take an extreme position. I will not, like D'Arcy Thompson and many of the purer "structuralists" who have followed him, take the position that those innate and internal properties of matter are the dominant forces to reckon with in the development and evolution of living form. Nor will I take the position that selection is all; rather, that there is a harmonious combination of the two. It has always struck me that the reasons for ruling out the physical forces in the development of form come from wearing one of a variety of blinders. For many molecular biologists it is simply a non-issue because for them the activity of genes is the be-all and end-all; genes have the plan for everything, including the form of the organism, and all other issues are beside the point or, even worse, not genuine hard science. Those who see and study the remarkable powers of natural selection worry mostly about small changes rather than the great ones of radical differences in form, and they therefore can think of the laws of physics only as something of little immediacy. At the other extreme, there are those who think there is some fundamental inadequacy in Darwinism. The great shapes in nature, and their fantastically complex organization, could not possibly be explained by something so utterly simplistic as selection. There are underlying laws and rules that can be perceived only through physics and mathematics, and they, as D'Arcy Thompson advocated so elegantly, will lead us to the truth. When one considers these two extremes, it seems to me the only possible answer is that both are right!

I do not wish to imply that this position of compromise is something that is in any way unusual. It is the position of many if not most biologists who have thought about the problem; those who take one of the extreme positions are relatively rare. It is true, however, that there is a human tendency to lean more on one side than the other—often the result of responding to a position held by someone else. Those who emphasize the importance of the role of the properties of matter in evolution are in part responding to the population geneticists who are concerned solely with the frequency of alleles in a population, and in part to the developmental geneticists who focus on the genes that are directly involved in the steps that govern development. By pointing out, quite correctly, that these ways of looking at population and developmental changes are only a portion of the larger problem of evolution, they place themselves in a strong position; but all too often they become carried away in their zeal, unearthing the errors of the ways of tradition and current fashion and overstating their case to drive their point home. It is a little like saying our heart is more important than our liver, when obviously we depend rather heavily on both.

There is a great rift between mathematically and non-mathematically minded biologists as to what comprises a satisfactory explanation. I belong more in the latter category and D'Arcy Thompson in the former—it is that aspect of *On Growth and Form* that has always fascinated me. The dualism no doubt goes back to early Greek philosophy (as one might expect from the classicist Thompson was) and is the modern incarnation of the rift between the idealistic thinking of Plato and the more observational thinking of Aristotle. Explanation is entwined with causation, and knowledge of the most immediate causes is totally satisfying to some, while others want to understand the ultimate causes and are little concerned with the details of how one gets there. To translate this into the terms of modern biologists, many of them get their entire satisfaction identifying the genes and the proteins they design, and learning the function of those proteins. The other great extreme is, what is the nature of the universe—is there a God that ordains its every detail? For a more biological version of the other pole, we have D'Arcy Thompson and his view that there are laws of nature that embody the principles of chemistry and physics, and that these laws can be described in the language of mathematics. Once this has been accomplished, the real meaning of the living phenomenon is revealed.

One wonders to what degree this schism correlates with mathematical ability. It is certainly true that one cannot completely embrace the mathematical extreme as being sufficient if one cannot manage much more than addition and subtraction. And obviously those who can see and master the beauty and power of mathematics might be expected to find in them the core of everything. Even those with mathematical disabilities can easily see what mathematics does and what it achieves; it is possible to have a reasonable grasp of statistics, or population genetics, even though one cannot derive any of the equations. This in itself is not the reason one requires a more immediate explanation of living phenomena. Here I am only asking if there is a correlation between the psychological need for a particular kind of explanation and the structure of the mathematical part of one's brain.

It has been brought home to me on numerous occasions that one of the things mathematicians find especially disturbing is the helter-skelter nature of mutation and natural selection. How can any organized plan be established by the simple process of encouraging or culling certain genes? They find this a particularly troublesome problem for the initiation or radical change of some major organ of the body during the course of evolution. The difficulty is that they do not appreciate the sequential and cumulative work of natural selection over many generations. I remember being asked by a distinguished mathematician if I did not share his view that Darwin had a trivial mind. When I said that I did not and explained why, he became very curious (and I was pleased to hear that later he spent much time reading about Darwinism). When I asked him what had led him to such a view, his answer was revealing. He said that he thought of the world, indeed the universe, as being the embodiment of perfection, and that it was his duty as a mathematician to try and understand that beauty. This being so, how could something as wonderfully organized as a living being be accounted for by the exceedingly messy and bumbling process of natural selection? Natural selection and the whole structure of mathematics and mathematical physics seem to be fundamentally at odds with each other. Of course they are not, for both exist, and the marriage of the two is what makes the problem especially interesting.

In recent years physicists and mathematicians have increasingly turned their attention to biological problems. Their skill in subduing some of the basic questions concerning matter and the cosmos have made it quite natural that they should examine what seem to be the intractable questions of the living. One of the issues that has become central to this new wave of interest in biology is that of complexity. There are a number of reasons for this, one being the increase in knowledge and understanding of complicated communication systems. This has led to a special interest in the working of the brain and the creating of computer analogs to achieve insight into our nervous system, and that of other animals. Another factor that draws our attention to complexity is the vast and obvious complexity of the environment. Great stress is being placed by ecologists on the immense number of species that exist in any one ecological community, and how those species interact with one another. This has led to a multitude of mathematical models, from those that apply to food chains to those that apply to stability and instability in ecosystems. The latter has led not only to a new interest in the mathematical properties of nonlinear equations, but to the application of these properties to weather forecasting and predicting economic trends.

Another direction has been a search for the nature of complexity that lies directly in living matter. One of the pioneers was Walter Elsasser (1966, 1975), who approached biological complexity with the tools of a theoretical physicist. He held that the principal difference between living and nonliving matter was that the former was made up of a myriad of different components—or was inhomogeneous, as he called it—and he strove to frame the properties of this inhomogeneity in mathematical terms. His work was no doubt before its time, and I no longer see any reference to it in the present-day literature. I suspect it might be worth going back to have a look at his ideas; when he was writing he seemed to be quite alone, but now there is a pack.

In recent years there has been a surge of interest in the role of self-organization in living systems. In some ways this new interest is a kind of sophisticated, modern structuralism. On purely theoretical grounds there has been a quest for the rules and properties that might underlie the way complex organisms are put together and that would account for many of the characteristics of life. For instance, a theoretical basis for a division of labor among units that coexist, such as cells, has been proposed, suggesting that a division of labor is inevitable for fundamental, innate reasons. The great problem mortal biologists such as myself have with these ideas is their very abstraction. I do not deny the strength of the mathematical argument, but I still do not understand how the biological result is achieved. I want to understand form, the division of labor, and biological organization in more concrete terms; to put it another way, I want the mathematics to be such that I can follow intuitively. I want to be able to grasp the point with both hands and have something solid to feel between my fingers. For me, all too often holding an abstraction is like trying to hold a fog. However, I am quite aware that this is a very personal view, and although it may be shared by nonmathematical biologists, it is no doubt shortsighted. It is easy to imagine that the day might soon come when these abstract analyses of self-organization may end up providing flashes of insight into the process of evolution that will prove not only to be important advances, but shed light that is clear to all.

Quite a different way of looking at the physico-chemical rules and life is the older idea that these rules set the limits of what selection can or cannot do. They do not in themselves create development or evolution; they just are ever present and allow no violations. They are sometimes called "constraints," although there is something about that term that bothers me. It seems to belittle the central role of all the rules laid down by the properties of matter. It would be foolish to try and put a measure on the relative size of the contribution of natural selection and the physical and chemical properties of organisms that lead to their evolution and development because, as I have said, each is both enormous and absolutely indispensable.

Let me now give some concrete examples that illustrate the dual roles of physics and selection in the production of form in organisms. I shall begin with venerable ones examined so elegantly by D'Arcy Thompson in his stellar chapter, "On Magnitude." For instance, there is the relationship between weight and strength. The central point is that with an increase in size, weight goes up as the cube of the linear dimensions (L^3), while strength only rises as the square of the linear dimensions (L^2). For this reason, if any structure is to remain sound as it becomes larger, it must, by various devices, keep its strength-to-weight ratio within reasonable and safe bounds. It is for this reason that big bridges are made of wires and lacelike struts—to keep the strength and minimize the weight.

One of the more obvious ways in which this dimensional problem may be seen is in the shape of trees that must be sufficiently strong to avoid buckling. The same situation applies to the limbs of animals where the weight they support becomes relatively large with an appreciable increase in size. Trees become progressively thicker with size increase, as we all know when we compare the thin sapling of an oak tree with the huge, thick tree it becomes when it matures. In quadrupeds, the legs of a small gazelle are delicate and slender by comparison to the massive, cumbersome limbs of an elephant. In each of these examples, there is a disproportionate thickness with size increase.

Experts are not in complete agreement as to what is the property which causes the thickness, with an increase in size, to be approximately proportional to the 3/2 power of the length ($D \propto L^{3/2}$). A most interesting hypothesis is that of T. A. McMahon (1973), who suggested that both limbs and tree trunks are so designed that their elastic properties re-

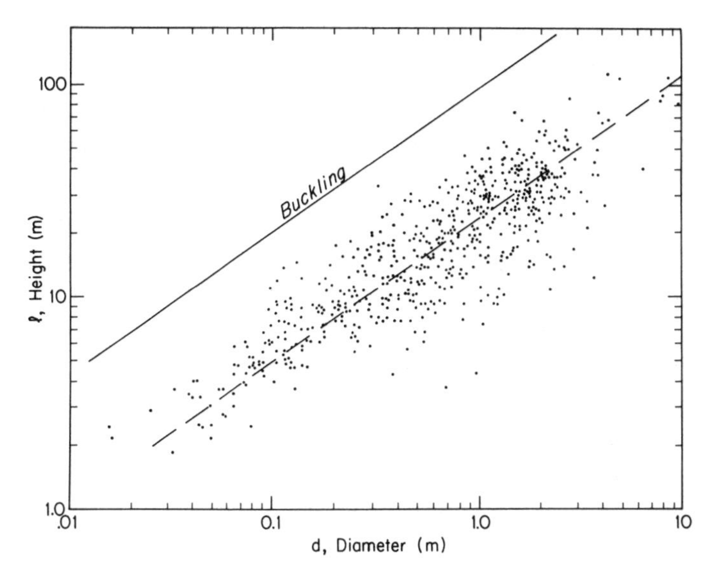

Fig. 1. Diameter of the base of trees plotted against their overall height on a logarithmic scale. The dots are 576 record specimen trees, representing what is believed to be the tallest and the broadest of each species found in the United States. (From McMahon 1973)

main the same regardless of the size. There is a sound theoretical basis for this, and if the notion is tested empirically, as McMahon has done with trees (among other organisms), it is clear that all the different size trees on a graph on which their length and their girth are plotted are safely below the buckling line (fig. 1).

We can now add the contribution of natural selection to this physical description of size and form in trees. If a tree were to appear with a mutant gene that causes it to grow disproportionately thin and as a result lie above the buckling line, the first great wind would blow it over and it would not survive to reproduce; there will be strong selection pressure for its elimination. There could be other genetic changes that might alter the structure of the wood and thereby affect its elasticity. Such changes in the properties of the wood might allow the tree to be thinner (or require it to be thicker), and this could account for some of the scatter of the points on figure 1.

Besides the relation between the weight and the strength of an organism and how it affects shape, there is also an important relation between the volume of an organism and its surface area. Again, one rises as the cube of the linear dimensions (L^3) and the other as the square (L^2). As has been pointed out by D'Arcy Thompson and many others, this is of special interest because some chemical activities, such as the overall metabolism, are a function of the volume, while diffusion of the gases involved in metabolism, such as oxygen and carbon dioxide, pass through the surface. It is quite obvious that with size increase, if the metabolism is to operate without difficulty, there will have to be a relative increase in surface area to keep pace with the increase in volume. One change that occurs is that the metabolism for a given volume of tissue goes down as an animal gets larger, but this compensating device is nowhere near sufficient to solve all the problems of size increase.

It is important to remember that the gases involved in metabolism are not the only substances that cross the surfaces in living organisms. In animals, food has to be assimilated in a gut and, as one might expect, the larger the animal the greater the proportion of gut surface. Small nematodes, for example, have a straight tube from one end of the body to the other, while mammals have an enormously convoluted gut with millions of small villi protruding into the lumen of the intestine that vastly increase the surface area and therefore the ability of the large animal to assimilate food. Again in this case the surface is trying to keep up with the volume as the overall size increases.

Perhaps the best example of all is the evolution of circulation. If a hypothetical spherical organism is a millimeter or less in size, the diffusion of chemicals, especially gases, is an easy matter, for in no time oxygen can reach any part of the organism, and carbon dioxide can easily escape. If the same-shaped organism were a centimeter or more in thickness, this would no longer be true. In fact, the cells at the center would soon be dead from lack of oxygen. There are two main solutions to the problem. One is to change the animal's shape by making it thin and flat, so that any cell is less than a millimeter from the surface; another solution, which is the only possible one for animals that are very much larger than a millimeter, is to devise a circulatory system so that oxygen can be brought to tissues deep inside and the carbon dioxide is removed. The ultimate in such a trend is what we find in vertebrates, where there is an elaborate circulatory system with a heart pump that pushes the blood to distribute food and oxygen to all the tissues into every inaccessible part of the body, and to carry off the wastes to be dumped outside.

The efficiency of these mechanical solutions to size increase is as impressive as their complexity—the question is, how did they arise? One might assume there has been a selection for size increase during the course of evolution, and that all the details of construction to accommodate this increase followed automatically. I prefer to think that selection not only culled for larger size, but also for all those engineering details that are needed if the size is to increase without mechanical failure. The physical rules are the bare bones of what is possible—selection may whittle away here and there, but those bones must remain intact.

Earlier I made the point that both physico-chemical forces (including self-organization) and natural selection played tremendous and indispensable roles in the production of biological form. Now we have gone one step further to make clear they are not the same thing—they are complementary. One does the picking and choosing of the genes, while the other sets the rule that determines which genes will be picked—that is, which genes can and which genes cannot make it to the next generation.

Let me illustrate the matter with a quite different example. In recent years there has been great interest, especially among some developmental biologists, in the use of reaction-diffusion models to describe patterns and how they might arise in development. The person who first put forth the idea in any useful form was the mathematician A. M. Turing (1952) in what has now become a famous paper. His basic idea is very simple. In a tissue there can be key substances that are given off by cells, and, depending upon the rate at which they diffuse and how they react chemically with one another or in themselves, they can produce a pattern, a shape. In the simplest case imagine a substance that is autocatalytic, that is, it stimulates its own production. Assume another substance that inhibits the production of the first substance. These are called the *activator* and the *inhibitor*, respectively. If the inhibitor is a smaller molecule than the activator and therefore diffuses more rapidly, on purely mathematical grounds Turing was able to show that these substances would form a gradient in a tissue. If these substances can actively affect the pattern in some way, such as by turning certain genes on or off, then they could play a key role in fashioning the form of the organism. For this reason Turing called these key substances "morphogens."

In recent years these ideas and their application to various living shapes and patterns have generated a large and interesting literature (see, for example, Meinhardt 1982; Segel 1984; Murray 1989; Edelstein-Keshet 1988; Cox 1992; Harrison 1993).

There have been a number of different applications of these reaction-diffusion equations to developmental systems. To give a few examples, they range from the control of the distribution of tentacles in hydroids (Turing's original example), to other aspects of their structure, to the proportions of spore and stalk cells in cellular slime molds, to the stripes and spots of zebras, leopards, and other cats, and to the initial laying down of pattern in the developing *Drosophila* embryo. In each case the theory provides a neat explanation of the pattern, and at the same time there is a rush of experimental work to try to identify the morphogens involved. This direct, experimental approach has turned out to be much more difficult, and the progress has been slow. Nevertheless, in some cases it has been both rewarding and profitable. The complication has often been that there are many more morphogens than are needed for the simple theory, but it is interesting to contemplate why this might be so. One possibility is that the extra morphogens fine-tune the regulation process so that there is less chance for deviation and error.

The reaction-diffusion model is a perfect example of a physico-chemical model of how form and pattern can be imposed on a living system. Let us now combine it with natural selection to see how the two could work together. Small differences in the receptor-binding affinities of the activator and inhibitor molecules would lead to major changes in the pattern, especially those that are near some threshold instability, such as the transition of stripes to spots in animal coloration. It is easy to conceive of selecting genes that affect the receptor proteins, and it would be equally easy to postulate gene changes that would produce small differences in the way the activators catalyze themselves. This is an excellent instance of how the physical and chemical properties generated by the molecules inside a living organism can work with natural selection to produce evolutionary change.

Another example of physico-chemical forces playing a role in the development of form by some sort of self-organization is seen in the consequences of the adhesive properties of the cell in a developing embryo. Much of the fundamental work is that of Steinberg (1970 *et seq.*), and Newman (1994) discusses this work along with reaction-diffusion phenomena to show how these properties of matter might play a role in evolution. Again these properties can be influenced by genes; for instance, a change in a gene may change the number of the adhesive molecules on the cell surface, or they may change the nature of the adhesive glycoprotein itself, but the physical and chemical properties are the given.

There is an interesting point, especially emphasized by Harrison (1993), concerning these mathematical models that describe form and the development of that form. There are two types: one is structural and the other kinetic, involving time. D'Arcy Thompson's mechanical and physical analyses of living form are classic examples of structural models. This is also the case for the molecular approach of the modern developmental geneticist, for there the essential element is the structure of the genes and their protein progeny, and how those proteins affect the emerging form. In contrast, the kinetic view involves time and concerns rates of processes such as diffusion and chemical reactions. Turing's reaction-diffusion models are an obvious example. To a mathematician, the difference between the structural and the kinetic models is great because the equations of the former are linear, while those of the latter enter into the more interesting realm of nonlinearity, where unpredictability and even chaos can hold sway. Another important difference is that in kinetic models the components are either changing or they are in a steady state, while in the structural models the end result is always a final equilibrium.

The developmental biologist is quite used to thinking in terms of time, for changes through time are the essence of development. As a result it is not difficult to see that a combination of both structural and kinetic processes are involved. It is tempting to think of all the extraordinary series of changes that take place during development as involving to a large degree (although by no means entirely) kinetic processes, and that equilibrium sets in as the form becomes adult. This may not be a true equilibrium; it could be that a persistent steady state simply becomes congealed and immobilized with maturity.

To prove my rule that the physical properties of living matter work hand in hand with natural selection, let me give a fascinating exception that comes directly from D'Arcy Thompson, from his discussion of the shapes of spicules and skeletons of radiolarians and foraminiferans. He makes two main points: one is that natural selection could not be operating, and the other is that while their fine structure may be crystalline, their external shape is not. His most elegant demonstration of this second point is by showing that in Ernst Haeckel's description of some four thousand species of radiolarians collected during the voyage of the H.M.S.

Challenger, two forms—one a regular dodecahedron (twelve equilateral pentagons) and one a regular icosohedron (twenty equilateral triangles)—are impossible crystal shapes (fig. 2): they violate the basic crystallographic law of rational indices. Crystals involve the stacking of identical molecules, and, like the piles of cannon balls by the village monument, they can be stacked in a finite number of ways. Haeckel's two species of radiolarians have shapes that do not comply with this rule—their symmetries are impossible ones for crystals. Haeckel concluded from his study of radiolarians and of the spicules of other animals that there was a formative element within these organisms which he called "biocrystallization." He postulated an inner force, a mystical structuralism that somehow directed the form of the organism.

D'Arcy Thompson did not subscribe to this view, nor did he feel that the great variety of these small-shelled protozoa could be accounted for by natural selection. Instead, he argued that they represent small variations in molecular forces such as one finds in the striking variation in the shapes of snow crystals, which is made possible by their small size. He is particularly eloquent in his dismissal of the ideas of men such as Rhumbler, who argued that the variation of the structure of the skeletons of these minute protozoa was due to a selection for strength and was

> the guiding principle in foraminiferal evolution, and marks the historic stages of their development in geologic time. But in days gone by I used to see the beach of a little Connemara bay bestrewn with millions upon millions of foraminiferal shells, simple Lagenae, less simple Nodosariae, more complex Rotaliae: all drifted

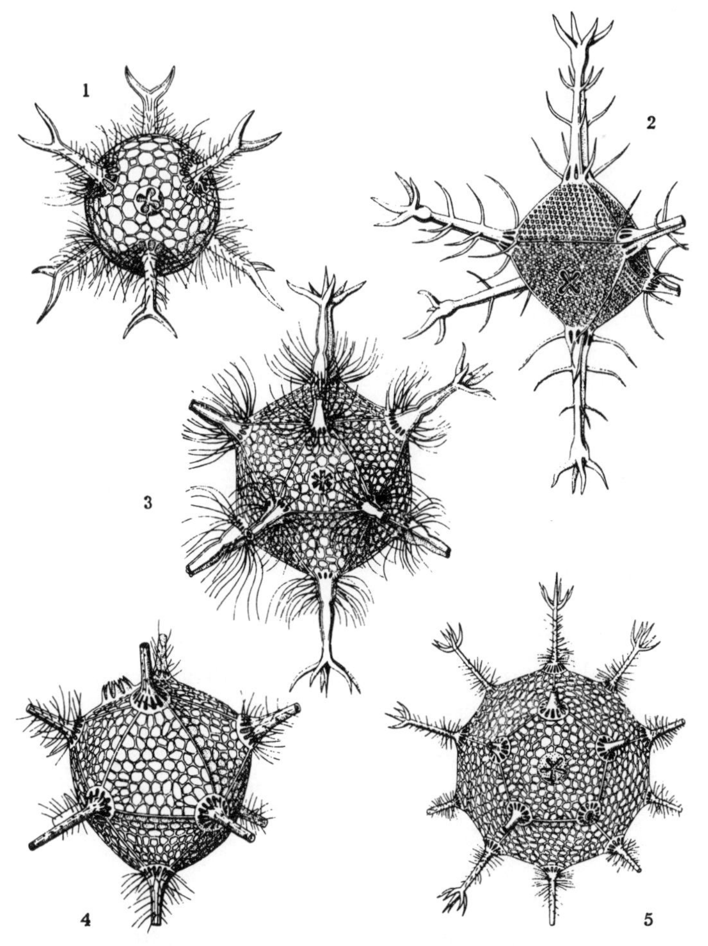

Fig. 2. Skeletons of various radiolarians, after Haeckel. (1) *Circoporus sexfurcus*; (2) *C. octahedrus*; (3) *Circogonia icosahedra*; (4) *Circospathis novena*; (5) *Circorrhegma dodecahedra*. (From Thompson 1942)

> by wave and gentle current from their sea-cradle to their sandy grave: all lying bleached and dead: one more delicate than another, but all (or vast multitudes of them) perfect and unbroken. And so I am not inclined to believe that niceties of form affect the case very much: nor in general that foraminiferal life involves a struggle for existence wherein breakage is a danger to be averted, and strength an advantage to be ensured.

His argument seems to me both sensible and convincing. How then are we to account for the great variety of shapes in these microscopic forms? D'Arcy Thompson's demonstrations that the shapes are not due to crystallization forces (of which I mentioned one) are equally convincing, and this means that, as he pointed out, there is something in the living substance that directs the shapes. Today we would say that there are specific proteins that guide the formation of the spicules, and since they are proteins they are under genetic control, which is an obvious conclusion since we know that generally the form characteristic of different species of foraminiferans and radiolarians is inherited. How then does one account for the enormously diverse and fanciful architecture of these microscopic creatures?

The most likely possibility is that this is an unusual example of neutral mutations—unusual because instead of the mutations being hidden in the genome as is ordinarily the case, many of the small DNA changes find their overt and direct expression in the variable shape of the skeletons. So here is an instance where the effect of selection appears to be totally absent, or at best minimal, and the physical forces hold full sway, except that they vary according to chance mutations that slightly alter the structure of the proteins responsible for the architecture of the shells.

What this tells us is that it is possible to have evolutionary change in biological form without natural selection. The multitude of species of radiolarians and foraminiferans is evidence that this might be so. Yet it must be true that selection was responsible for producing the first protozoan and the first representative species of these two major groups of protozoans. It is only after this stage that selection subsided and neutral mutation, with the molecular variety it produced, took over. I put this forth as the exception that proves the rule, the rule being that usually natural selection is constantly pecking at the gene changes that appear, some being favored and others eliminated in successive generations. If one withholds selection, as may have happened for the shells of these protozoa, then the physical forces are liberated and can show seemingly infinite variations. However, even in this case there are bound to be limits on what physical and chemical constructs are possible if the organisms are to exist at all.

CHAPTER 2

Competition and Natural Selection

BECAUSE of the very nature of natural selection, with its picking and choosing for reproductive success at each generation, one would expect that all organisms that have been subjected to this rigid test would automatically have a conspicuous efficiency to all their activities. After all, being selected are those properties that will survive over a whole series of generations. There is a selection for the efficient handling of the machinery within an organism, its development and its physiology, as well as the efficient handling of the relation of that organism to its environment. This is our standard understanding of natural selection.

Here I am interested in the general form this efficiency takes, and it is clear that at all levels it produces a kind of order that is hierarchical. There are generally a series of levels, each one holding sway over the one below it, and the arrangement of each level, from molecular behavior to animal behavior, is in a chain of command. Even the parts of any one level are competing with one another, and some will dominate others as well as those at the level below. It is a complex network of interacting components in which the overriding principle of competition takes many forms. The term "competition" is a convenient shorthand way of referring to the order that is the result of the sieve of natural selection.

I am well aware that my cavalier use of this term will offend those who revel in precise definitions. Furthermore, I will use the words "dominance," "peck order," and "chain of command" as obvious manifestations of competition. The reason for putting them all in the same basket is that the exact way selection produces order differs from case to case, and any one of these terms covers some or all of the properties that are so characteristic of living organisms. To see the point, consider the fact that often selection favors the cooperative and symbiotic associations that are so common in nature, yet they are nothing more than another facet of competition. It is a case of "if you can't lick 'em, join 'em," for by means of the association one or both of the organisms may be able to improve their chances of reproductive success, so that a special case of competition becomes winning by joining.

Competition, including competition by association, occurs at all levels, which is to be expected because selection operates at all levels. Here I will systematically move from the smallest, most primitive level involving competition among the genes, to the other end of the spectrum where there is competition in the behavior of animals, even to the extent that the human mind inserts the elements of competition into the inventions it creates. I will show how at each level natural selection has made itself evident in the form of different kinds of competition. Different they may be, but all are ultimately represented by genes. As we shall see, this competition has led, through the course of evolution, to the most extraordinary secondary and tertiary forms of competition.

The Molecular Level

The easiest way of examining competition at the molecular level is to consider the steps that might have been part of the origin of life. It is generally thought today that the first step in evolution towards living cells was a system of molecules capable of polymerizing into chains that had the power of replicating themselves by making templates. One possibility is that this initial polymer was RNA, and there is considerable evidence to support the idea that the reactions involved—the polymerization and the replication—were originally catalyzed by RNA itself.

RNA replication involves the repeated formation of templates. One might ask, why would a molecule perpetuate itself at all? The answer is that once a chemical system appears that forms templates, the template is made automatically if it is surrounded by the right building blocks; since the new molecule has the same property, it in turn will form templates. Therefore, if the ability to form templates occurs—presumably by pure accident—cycling will occur by continual and repeated template formation from then on. This is the beginning of cycling, and the cycling will be nonstop and endless (provided it does not run out of the necessary ingredients).

Cycling is also the beginning of natural selection, because there will always be the possibility that some variant of the template-making polymer will outcompete its progenitor. The competition will be between those polymers that polymerize and replicate more rapidly and more reliably than others; they will be the winners, the survivors, because they have been more efficient in grabbing the bases, the building blocks, that are available. As a result they become the progenitors of future molecules of a similar sort. Those successes mean a selection for better catalysts and for better control of the whole process. All the changes arise by selection; natural selection inevitably turns variation into competition.

The conventional hypothesis is that this selection ultimately led to the appearance of DNA, which forms an RNA template, which in turn produces protein. In that one short sentence I have rushed over an evolutionary process that must have taken an exceedingly long period of geological time, and there is no need, for my purposes, to speculate on the nature and sequence of the innumerable steps that led to this ultimate result.

The competition is for capturing the energy. In order to make this objective more successful, two elements arose—two that go hand in hand. One is the production of some mechanism, involving protein enzymes, whereby energy could be harnessed and processed to run the replication in a reliable and controlled way. Undoubtedly that would not be possible without the development of a way to enclose the replicating machine and the newly evolved metabolic machine, a feat that was accomplished by the production of a membrane. The result was the formation of the first cell.

However, evolution did not stop with the formation of a cell—there is a vast range in the size and complexity of organisms living today. This seems contrary to the idea that the smallest and most rapid and efficient cyclic replicators should always be the winners, for if this were true one would expect that the world would be made up of nothing but minute bacteria. The explanation for the existing diversity is, of course, well understood; the environment is complex and varied, so that there is not just one competition for one niche, but many, and each niche can be filled with a different kind of organism. Inevitably these will vary enormously in their size and complexity.

It is necessary to make clear that the idea of competition automatically encompasses the notion of dominance. If there is a competition, then there is a winner and a loser, and the winner becomes dominant over the loser. This rather obvious point is illustrated over and over again if one examines how the various components within a living system interact with one another as they compete.

The idea that one cycling, replicating polymer may be more successful than another means that in the race for populating the world with its issue, one polymer, by a difference in its chemical composition, becomes dominant over another in terms of replicating success. More important is the fact that as soon as catalysis plays a part—even the primitive catalysis of RNA controlling its own replicating and polymerizing activities—the catalyst dominates over the replicating RNA in the sense that it controls its own cyclic changes. We then have one molecule controlling another, and the dominant molecule, should it vary in any way, will be subject to natural selection. The more effective variant will replace the less efficient one.

The most important level in this early evolution, which must have involved innumerable steps, is the appearance of the DNA, RNA, protein sequence. By now there were a great multitude of enzymes which controlled every chemical move in the elaborate and involved progression. With such a huge battery of enzymes it is no longer meaningful to say that each one dominates or controls a particular chemical reaction; now the great mass of those reactions must be organized in such a way that the process works as a whole, as a unit.

One way this could happen is by a time sequence, where B can occur only if A has taken place, and C can occur only following B, and so forth. By having such a time-dependent sequence one can produce order within a multitude of smaller reactions. They become interrelated and ordered solely by virtue of their locked-in sequence.

However, this is by no means the only form of an early selection for complex, interrelated chemical reactions. The products of one reaction can inhibit or stimulate some other reaction in the chain and increase or decrease its activity. There are many examples of feed-back inhibition in which a whole sequence of reactions may be controlled. For instance, in a linear chain of reactions from A to G, the final product G might inhibit the reaction from A to B. In this way there is always a controlled amount of the final product G. In the case of gene action, often the genes are prevented from being active at all by the presence of a specific protein that is a repressor which sits on the gene and prevents its transcription. This in turn can be removed by another protein (a derepressor). This latter case is one of activation, and there are many such instances in the process of protein synthesis, including catalysts which promote as well as control specific reactions. Therefore random errors in DNA structure and selection have produced a complex of inhibitors and activators which see to it that the elaborate chain of chemical reactions involved in the synthesis of specific proteins are under strict control. Each innovation in the chemical pathway has been tested by natural selection; if it does not increase or at least maintain efficiency, in the sense of reproductive or replicative success, then it is immediately eliminated. On the other hand, if it is a favorable mutation, it will be retained. There is no way any change that produces disorganization or inefficiency can survive the scrutiny of natural selection in the successive replication cycles. It is required that successful changes be integrated and that they increase the cycling reliability.

For the control to be as effective as possible it must be centralized. It started off centralized because in the beginning presumably there was only one thing: cycling nucleic acid. First it only altered itself, but eventually any change in the DNA automatically meant a change in the protein. One assumes that initially these proteins were mainly concerned with the immediate replication process, but eventually they could assume other useful functions. These involved the formation of the metabolic machine, which meant in particular the production of proteins that could catalyze reactions that would harness energy for maintenance and for the replication process itself. One of the astounding things is how many proteins and how many genes this involved. Furthermore, not only must the proteins be produced in the right quantity, but they also must be in the right place. The chain of command begins with the nucleic acid and leads to the many steps that follow.

The Level of the Cell

The notion of the spacing of the key molecules brings us to a new function of selection which ultimately involves the construction of a cell. In the smallest prokaryotes, which we assume were the first kinds of cells, there is relatively little in the way of compartments. The DNA consists of one circular strand which lies near the center of the cell, and the other major structure is the cell membrane and associated cell wall. Many of the proteins are free in the cytoplasm, while others are lodged in the lipid plasma membrane. The sources of energy for bacteria may be remarkably varied: they may obtain it by promoting simple, energy-liberating chemical reactions in their immediate environment; they may obtain it by taking in organic substances and get the energy from their breakdown by controlled oxidation; in some bacteria, especially cyanobacteria, they convert the energy from the sun. In the latter case these photosynthetic bacteria will have special internal membranous bodies that contain the pigment which absorbs the light energy. This energy is imparted to the catalytic proteins that synthesize all the molecules needed for their growth and reproduction. In all these instances, all the proteins needed for energy conversion are made by instructions from the strand of DNA. Once the proteins are formed and begin to perform the function for which they are fitted, they are independent of the DNA. However, their activities are not anarchic: they are selected to produce sequence-dependent chains of reactions with the capability of feedback inhibition and self-activation (autocatalysis), all of which produces an orderly activity, from the initial capturing of energy to the use of that energy for maintenance and the growth of the cell.

The next big step in the evolution of the cell is the appearance of eukaryotic cells. For many years it has been argued convincingly that some of the organelles arose through the symbiotic association of prokaryotic cells (for a review, see Sapp 1994). What is certain is that the mitochondria of eukaryotic cells (and the plastids of photosynthetic plant cells) closely resemble existing prokaryotes, and it is known that bacteria live inside many animal cells. The obvious suggestion is that an extended period of symbiosis led to a more permanent mutual dependence so that mitochondria and plastids of organisms today cannot live when isolated from their parent cell, nor can the parent cell live without them. This is invariably true of eukaryotic cells deprived of their mitochondria, but in photosynthetic flagellates such as *Euglena,* which can get their energy from the sun or organic matter, the removal of the plastids does not mean death if they are provided with suitable nutrients.

I shall return to mitochondria and plastids in a moment, but first I want to mention that there are numerous other innovations in a eukaryotic cell that are absent in prokaryotes. In the first place, there is a nucleus surrounded by a membrane. The DNA is complexed with various proteins, and eukaryotes have made the remarkable step of having chromosomes capable of mitosis and meiosis; replication has become a much more complex and organized process. This involves centrosomes and the formation of a spindle whose fibers separate the members of each pair of duplicated chromosomes. In the cytoplasm there are equally significant innovations, such as a variety of important membranes which form the endoplasmic reticulum and the Golgi apparatus and those that surround various kinds of vacuoles. The eukaryotic cell is a small factory and each of these structures is a compartment that carries out a specific function, such as synthesizing particular molecules and passing molecules—even large proteins—from one part of the cell to another. Besides these structures, which are to a large extent part of the metabolic machinery, there are fibers and associated bodies which involve structural proteins that are responsible for the shape of the cell and its movement.

The importance of all these cell parts from my point of view is that they are not a collection of structures rattling around in a bag, but they are beautifully organized so that the cell can function as a well-coordinated unit. They show a division of labor: some parts, such as the mitochondria, are responsible for energy entrapment and conversion of that energy into useful forms for the cell; others, such a the endoplasmic reticulum and its inclusions, are responsible for protein synthesis; still others, such as the various kinds of vacuoles, are involved in the transport and storage of specific molecules; the Golgi apparatus is responsible for the synthesis of molecules which are cell products to be secreted; finally, others are involved in various forms of locomotion—the function of the collections of structural proteins that form spindle fibers, flagella, and so forth. A eukaryotic cell is a remarkably well-organized, miniature factory. The fact that in general eukaryotic cells are larger than prokaryotic cells is undoubtedly in part responsible for this well-structured division of labor.

Let us now consider the chain of command within all this division of labor. There are enough separate components within a cell, each with its myriad of different molecules, to see that all these parts cannot act independently but must be coordinated in some way so that the whole assembly does not disintegrate into chaos. Obviously only those arrangements that are orderly and efficient will survive natural selection.

The most conspicuous evidence for competition comes from the dominance of the nucleus over the rest of the cell, largely because the directions for what proteins are to be synthesized come inevitably from the genes. Yet clearly there is rigid control for what genes are to be transcribed, and when that is to happen. This is achieved by elaborate feedback control pathways so that no protein is synthesized when it is not needed; when synthesis does occur, the amount must be carefully governed. Part of this feedback control takes place within the nucleus itself, and some genes can produce repressors or derepressors that orchestrate the sequence of events. In other words, some genes rule over others, and those ruling genes are the regulatory genes. There is no one gene that is king of all others; but there are a number of regulatory genes, and they are triggered by events that occur within the cell or in the environment. For instance, in the cell cycle there are a series of temporal changes, one dependent upon another, and these internal events activate a sequence of regulatory genes. So here is another example where competitive order is generated by a rigid sequence of chemical events that has arisen through competition.

Numerous external cues also play a part. If a cell is in an environment that contains a sugar for which it does not have the enzyme necessary to break it down for energy—yet it has the gene for the that enzyme—then the sugar will induce the dormant gene to start making the enzyme. In this example one can see how the environment can influence the genes to produce the right proteins.

Therefore, while the nucleus appears to be the dominant structure in the activities of the cell, it is equally clear that it does not dominate by giving off one signal but by a whole series of signals that are responsive to the internal and external conditions of the cell. This very sensitive flexibility makes the nucleus all the more successful in dominating and orchestrating the complex activities of a cell, a condition that was obviously encouraged by natural selection.

There is a particularity interesting aspect of nuclear dominance. As mentioned earlier, mitochondria have a prokaryotic-like ring of DNA—what is its relation to the DNA of the nucleus? The mitochondrial DNA is responsible for making very few proteins; the majority of the proteins that make up mitochondria are manufactured in the nucleus and transported to the mitochondria. The nucleus has clearly exerted its dominance over the ancestral symbiotic mitochondria.

This nuclear dominance is illustrated in a particularly interesting experiment of Thorsness and Fox (1990), who put a genetic marker on a plasmid that they could incorporate into either the nuclear or the mitochondrial DNA. Then they selected for its migration and showed that it was a relatively easy matter for the marker to move from the mitochondria to the nucleus, but it is at least 100,000 times less likely for the marker to move in the opposite direction. In other words, the nucleus exerts its dominance over the organelles in still another way: by controlling the direction of the traffic in wandering, gene-containing plasmids.

One of the most significant results of early natural selection was the emergence of sexuality. This means of reproduction provided a means of shuffling the genes so that the variation needed for natural selection could be sustained at the right level for optimal selection and maintenance of the most successful variants. That sex is ubiquitous in all groups of organisms living today is testimony to its success; we presume that it has been retained basically unaltered since early life on earth.

In bacteria, sex consists of the transfer of DNA fragments from one individual to another. This emphasizes the fragments' interesting ability to move, not only from one part of a strand to another but, in the case of sex in bacteria, from one cell to another. Presumably it is a fortuitous property of nucleic acid chains that has played a very significant role in the generation of variation and therefore of evolution. It is a property that not only permits variation in prokaryotes, but also enables crossing over in eukaryotic meiosis. It also allows the variation-producing effects of transposable elements (jumping genes) discovered by Barbara McClintock long ago, even before it was known that genes were DNA and that DNA strands were capable of being disrupted by insertions or deletions.

A particularly important step in early evolution must have been the appearance of chromosomes and their behavior in mitosis and meiosis in eukaryotic cells. It must have been an extraordinarily complicated evolution because so many new structures and new processes were involved, such as the construction of the chromosomes, the mechanism of their duplication, and their separation so that each daughter cell received an equal component of genes. Furthermore, it provided a mechanism by which two cells of different parents could fuse, giving the resulting zygote the complete gene complements from both parents. This meant that for some period a single cell would contain a double set of chromosomes: diploidy was invented. Diploidy provides many advantages such as the ability to hold a greater variety of alleles in one organism, and for this reason it is not surprising that in most complex organisms the diploid stage became the dominant phase of the life cycle. When one thinks of the many innovations eukaryotic sex involved—fertilization, meiosis with the shuffling of chromosomes and the crossing over that permits the shuffling of parts of the chromosomes, the duplication of the DNA strands without genetic loss, and the extraordinary number of catalyzed chemical reactions needed for these processes—it was indeed an extraordinary achievement.

One might argue that because the whole system of eukaryotic sex is so complex, how could it have evolved by small steps? But this problem is no greater than the old one of the evolution of the vertebrate eye, where we now accept the view, first proposed by Darwin, that many intermediate, less elaborate steps are possible; the eye does not present a problem for the evolution of a complex structure. Just as the eye can be traced back to primitive photoreceptors, no doubt sex and its attendant cellular phenomena can be traced back to transformation in bacteria. The big difference is that in the case of the cellular mechanisms of sexual reproduction, no intermediate steps are present today, which might give the false impression that it arose full-blown by a miracle from outer space. A more rational argument would be that the final product, as we know it today, was so much more successful than its prior, intermediate evolutionary steps that those steps were eliminated by competition many millions of years ago. (See Maynard Smith and Szathmáry 1995 for an interesting review of these questions.)

Even more remarkable is the fact that many aspects of the cellular sexual system seem to have remained essentially unchanged once it appeared in a primeval eukaryote. This can only mean that it has been supremely successful as a mechanism for storing and shuffling variation, so much so that it is firmly kept in place by selection because it is a wonderfully effective vehicle for the promotion of selection in animals and plants. The sexual system is a double-edged sword: it promotes itself, and it provides a perfect device for all the variation needed for the evolution of all the other features of organisms. The fact that some organisms are asexual for some of their life cycle, or some organisms seem to be permanently asexual, is in no way an impediment to the argument for this success story. When environmental conditions remain unchanged for extensive periods, asexual or clonal reproduction is the fastest way to reproduce. If external conditions become altered and unfavorable for the asexual phenotype, it can either go into a sexual phase, or, if it has lost this ability through continued environmental constancy, it will go extinct. The environmental fluctuations keep sexuality in place, and all the worries about it being more expensive than asexual reproduction can be cast aside, for sex is well worth the price. There are, of course, other arguments for the advantages of sex which are undoubtedly also true (see reviews in Bell 1982; Michod and Levin 1988).

Let me now consider the question of dominance in the cellular mechanism of the sexual process. I have already stressed the sequence order in the mitotic cell cycle, and presumably that basic scheme also applies to meiosis, although there are obviously many additional features in the process of reduction division. But in general the control mechanism must again be a sequential one. This would apply to the process of fertilization and the pairing of the chromosomes of the two parents. Dominance does not seem to be an obvious factor in these cyclic, cellular phenomena, but it becomes a very significant one in the diploid genome. This was first appreciated by Mendel even without knowledge of cell mechanics, and he demonstrated that inherited traits could be dominant or recessive. Now we know that there is competition between the two alleles at one locus, and that the dominant gene will be the successful one and be transcribed and translated into its protein. This is obviously something that is only possible in a diploid organism. It reflects the fact that when alleles are not identical, one will very likely lord it over the other. There are some interesting hypotheses as to how genetic dominance might have been produced by selection, but they go beyond the needs of our argument here; the relevant point for this discussion is the simple fact that this is one of a number of instances where genes show evidence of competition for expression.

Recently there has been a renewed interest by Hickey and Rose (1988), Hurst (1992), and others in the possibility that many of the aspects of sexuality that are observed in the organisms living today can be understood in terms of competition between the genes in the cytoplasm (in the form of plasmids) and those in the nucleus. If all of these genes are competing for their own interests—that is, for their own perpetuation—then a number of consequences are predicted. One of them is that this might be the cause and route of the origin of sex itself. If a plasmid gene appeared that was able to find a way to advance itself by fusing with and infecting other cells, it would dominate the cytoplasmic genes of the new host cells. The result would be two populations of cells: those with and those without the selfish plasmid gene. There would now be a struggle between the genes of the genome and the new plasmid genes, and the nuclear genes would have to find a way to control the parasitic plasmid genes. One way to do this is for uninfected cells to devise a way to prevent any new plasmid genes from entering the cytoplasm when fusion occurs. This is exactly what happens in the process of fertilization in most organisms. The cytoplasm for the zygote is mostly carried in the large egg; the sperm sheds its cytoplasm usually before it matures, and even its mitochondria normally are not passed on to the next generation. Sexuality then, in this hypothesis, is the direct result of trying to block the invasion of a selfish plasmid gene; fertilization becomes a way of blocking any foreign cytoplasmic gene from entering the uncontaminated maternal cytoplasm.

Going one step further, these workers have also made the suggestion that such gene competition is responsible for the evolution of anisogamy, where the egg is large and the sperm is small, with little or no cytoplasm. The argument is that this is a way of totally blocking any cytoplasmic genes from entering the maternal cytoplasm. In ancestral forms where the gametes are identical, or isogamous, there are other known mechanisms where the cytoplasmic genes of only one of the parents will prevail; the evolution of a large egg and a small cytoplasm-reduced sperm is simply another way of achieving the same thing. (There are other equally convincing arguments for the evolution of anisogamy, as, for example, that of Cox and Sethian 1985 and others. However, these various hypotheses are generally not mutually exclusive—all could be correct.)

One thing is clear: there is competition among the various genes within the cells of an organism which we can trace from the very beginning of replication of template molecules, right up to the competition of those molecules within a cell. That competition, which is run by natural selection, might even be responsible for the origin of sexuality.

The Multicellular Level

Moving one step up the evolutionary scale to multicellular organisms, competition and dominance are again conspicuous features. This is a venerable subject that has an interesting history in the the study of the development of organisms. It was the brain child of C. M. Child, a heretical embryologist at the University of Chicago who published extensively in the early part of this century and summarized his life work in one monumental volume (Child 1941). Early in his experiments he noted that in embryos and in developmental systems of all kinds, both animal and plant, there was a gradient of metabolic activity. First he showed this by using metabolic poisons, such as cyanide, and noted that invariably one particular part of the embryo (usually the anterior region) disintegrated first. After being much criticized for using such deadly indicators of living activity, he repeated the same experiments using oxidation-reduction vital dyes and confirmed the ubiquity of metabolic gradients.

From these observations he generalized that one part of a developing system became dominant over another in terms of metabolic activity by being more effective in grabbing the substrates needed for energy conversion, and once a dominant region was established it retained its position due to its success in competition for oxygen and food. He argued that the dominance system produced a gradient, but that the gradient could only extend a finite distance—there was a limit to the amount the dominance could control. Cells beyond the limit would set up a new dominance-gradient system. He called this "physiological isolation." From these observations he suggested that small quantitative differences in metabolic activity along the gradient led to qualitative differences, and for this reason metabolic gradients were the basis of all pattern. Here is where he met vigorous opposition. Many felt that a development pattern was often so complex that it was absurd to attribute it to simple gradients of oxygen consumption. Unfortunately, during his lifetime he was thrown out with the bathwater. Today we are able to free ourselves from the argument and from his strong advocacy. By being selective among his ideas we can now see that in some ways he was ahead of his time. Gradients and the principle of dominance are central to our modern view of how development works.

It may be helpful to look at an example which illustrates some of the properties of metabolic gradients and associated phenomena, for they really do exist and are of exceptional interest. A classic example is that of the colonial hydroid *Tubularia* (fig. 3). In common with hydroids in general, it consists of a branching tube that periodically gives rise to a feeding polyp capped by the mouth region, or hydranth. The tube has a lumen that is continuous so that all the polyps are connected through a common gastrovascular canal. The tissue itself consists of two conspicuous layers, an endoderm and an ectoderm, separated by a thin mesoglea.

Because *Tubularia* has a very tall stem bearing the hydranth, a number of workers found that it was a most favorable material for the study of regeneration. If small segments of the long stem are cut, each piece regenerates a new hydranth. Furthermore, the pieces have a polarity, for the new hydranth invariably forms at the distal, or anterior, end of the cut piece.

The first relevant experiment was performed by T. H. Morgan before his incarnation as a geneticist. (For reviews on all the work on *Tubularia*, see Barth 1940; Tardent 1963; Rose 1970.) He showed that the polarity could be reversed by sticking one end of a piece of stem in the sand: no matter which end is buried, even if it is the anterior end, the free end produces a hydranth. The explanation of this result came from Child and his students. The buried end has less access to oxygen and therefore loses its dominance to what had previously been the posterior end, which is now sticking up into the well-aerated seawater. Accordingly its metabolic rate becomes greater than that of the buried end, and as a result it becomes the new, dominant anterior end. Many variations of this experiment can be performed. For instance, it is possible to put small glass caps over one end of the stem, which will also reverse the polarity.

As certain as this story seemed, an opposing view was championed by S. M. Rose and others (see review in Rose 1970), who argued that an inhibitor is produced by the stem. If the inhibitor is not allowed to diffuse away from the cut end of a piece of stem, it prevents the regeneration at that end and keeps it from becoming dominant. To shorten a very lively controversy, it turned out that both camps were correct—either oxygen or an inhibitor could determine which end was to become dominant. The idea that an inhibitor might be involved is quite consistent with the widespread evidence that between-cell signals play an important role in development. In either case, competition is evident, for one end becomes dominant over the other.

Yet the metabolic gradients are real. This was shown in a straightforward experiment in which a stem was cut into small segments of equal size and each was put in a respirator. The anterior-most piece respired at the highest rate, the next one slightly less, and so forth down the line—a clear gradient of oxygen consumption was demonstrated. It was also possible to provide direct evidence for Child's physiological isolation. If a series of segments of different lengths is cut and allowed to regenerate, there is a certain length above which a new hydranth will appear at both ends (fig. 4). The distance from the anterior end to the posterior end is too great for the power of the anterior, dominant end to prevent the posterior end from producing a new hydranth. This

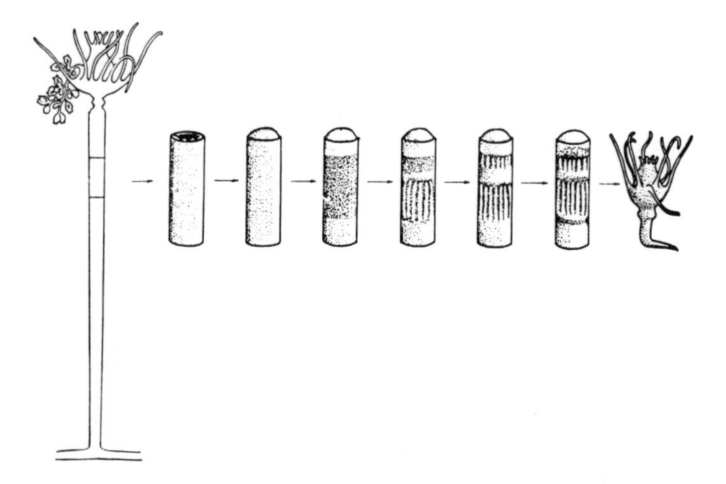

Fig. 3. An adult polyp of *Tubularia* and the steps in the regeneration of a cut portion of its stem. (From Tardent 1963)

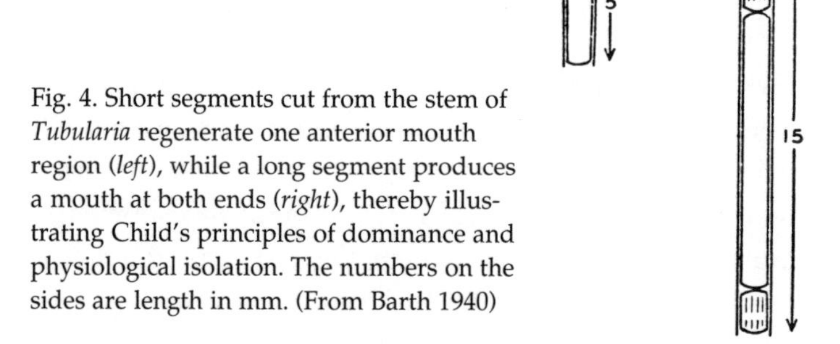

Fig. 4. Short segments cut from the stem of *Tubularia* regenerate one anterior mouth region (*left*), while a long segment produces a mouth at both ends (*right*), thereby illustrating Child's principles of dominance and physiological isolation. The numbers on the sides are length in mm. (From Barth 1940)

inhibition could have been the direct result of the inhibitor produced by the hydranth end that could only diffuse a certain distance, or the substrates necessary for high metabolism could not be reached at such a great distance, or both mechanisms could have operated together. *Tubularia* has the great advantage of illustrating, in a very clear manner, dominance, physiological isolation, metabolic gradients, and the role of between-cell inhibitors.

The Level of Behavior

In the evolution of animals an extraordinarily important new phenomenon arose that we call behavior. In the most primitive invertebrates the nervous system is involved solely in the simplest quick responses to the immediate environment, which are essentially reflexes. They consist of such actions as the avoidance of inimical directions in their immediate surroundings, the orientation towards (or away) from light, and orientation towards food. With increasing refinement of these primitive reactions, the nervous system evolved into something elaborate, producing progressively more sophisticated behaviors, the ultimate being our own remarkable powers of reasoning.

One of the key elements of behavior, beginning from its earliest appearance in evolution, is the ability to transfer information from one individual to another, and each individual organism can be affected by, or respond to, this information. This is similar to what is found in the development of multicellular organisms, where there is a production of chemical signals in one part of the embryo that affects another; such signals play a key role in coordinating and unifying the development of an embryo. In behavior the signals may also be chemical, but the whole process of sending and receiving information by one individual from another involves the entire nervous system, including the brain. As a result, the kind of information passed on can be far more elaborate than simple activation or inhibition (on or off)—in its ultimate form it can involve language and symbols so that an extraordinary amount of information can be transmitted in behavior.

The significant point for my purposes here is that the transmission of information is *not* genetic, but is on a different level. Much to the distress of some anthropologists, I have previously (1980) called change that emerged from this kind of information transmission "cultural," as opposed to "genetic." The distinction between the two is extremely important because of the way in which the information is transmitted. In genetic evolution all the information is passed solely from parent to offspring through the egg and sperm; there is no shortcut. This means that in order to introduce a new mutant gene in a population, many generations must be involved. Genes are the sole agency of change for pure Darwinian natural selection. Therefore the use of the word "natural" here denotes that one is selecting for genes, and nothing else.

Passing information by behavior has entirely different dynamics. R. Dawkins (1976) invented the useful word "meme" to signify any bit of behavioral information, be it an idea, a fad, or even a rumor. Compared to genes, memes can be passed from one individual to many, and this can occur very rapidly, with an immediate impact. Some memes may be slower to modify, as is the case with customs, but even long-standing customs can be quickly modified or eliminated. As is quite obvious, there is a radical difference between the modes of inheritance of genes and memes. Equally obvious is the fact that both can be selected, but the selection of memes is clearly less rigid, and their response to selection is far more immediate than what is involved in the natural selection of genes.

We must remember that the world of memes is not totally divorced from any direct influence of genes. Memes are a product of the nervous system and the brain, and their construction is the result of gene changes encouraged by natural selection over vast periods of time. No doubt this selection was fostered because of the remarkable ability of the nervous system to produce and receive memes.

One of the interesting things about behavioral transmission when compared to gene transmission is that it is not only more rapid, but far more flexible. Initially, in animals with the simplest of nervous systems, there is only one response to a signal—it is a single action response, an automatic reflex. The moth flies to a candle, the spider attacks when it feels vibrations in the web, and so forth. However, during the course of evolution there arose the ability to make choices—to choose between two or more behaviors. This is evident even among invertebrates, for one of the most obvious examples is mate choice in sexual selection. Usually it is the female that chooses among a number of males which, following Darwin, results in the disparity between the appearance of the sexes, from stag beetles to peacocks and peahens. Choices are not confined to mate selection, but also occur when an animal looks for prey or a nest site, decides on an escape plan from a predator, and while it carries out innumerable other tasks. I have called this kind of behavior "multiple choice behavior" (1980). Competition in behavior arises from this multiple choice behavior: it is the flexibility afforded by multiple choice behavior that facilitates competition among any group of interacting individuals.

The most important and fundamental manifestation of behavioral competition may be seen in the well-known phenomenon of peck order, first discovered in hens by T. Schjelderup-Ebbe in the early part of this century (see review in Brown 1975). If a group of hens is put together in a yard, very rapidly a dominance hierarchy will become established in which hen A will peck all the other hens; B all of them except A; C all except A and B, and so forth down the line. It was realized that these differences in behavior are not primarily genetic differences, for they could be modified by many factors, such as the degree of starvation, previous experience, age; and these and similar factors influence the hormonal state in the body and the brain. In other words, slight differences in internal chemistry will lead to a sorting out of individuals in a group into a dominance hierarchy. There is never an egalitarian group, but always a hierarchical one. Presumably this would be so even if all the individuals were genetically identical. This can be seen in human identical twins, where one individual in the pair will show subtle dominance over the other.

One tends, in our egocentric way, to think that only higher animals such as vertebrates are capable of multiple-choice behavior and show competition in the form of a peck order. This is clearly not the case, and, for example, among insects there are well-defined dominance systems, as is evident in social wasps, and even among the workers of some species of ants. Therefore, besides exhibiting multiple choice in sexual selection, insects are capable of a peck order as well.

One of the more interesting aspects of dominance in animals is that in species with large brains there is sometimes the formation of alliances between individuals to assure dominance. This was vividly demonstrated by F. de Waal (1982) in his extended studies of a group of chimpanzees in the Arnheim Zoo in the Netherlands. The group had three strong males who were seemingly capable of assuming the role of the alpha male. The only way one of them could achieve this top status was to solicit the help of one of the other males, and then the two could overpower the third. These maneuverings for alliances were aptly called "chimpanzee politics" by de Waal. More recently even more elaborate alliances were shown by R. Conner and his coworkers (1992) to exist in bottlenose dolphins. The males go about in small groups of two or three individuals, and if one such group is herding a receptive female, a rival group may solicit the help of a third group to capture the female. Here there is a two-tiered system of dominance alliances which seems to be the ultimate in sophistication. (Note that the forming of these alliances is an example of behavioral symbiosis, one of many instances where behavioral cooperation mimics the kind of symbiosis one finds within and between cells.) Dolphins, chimpanzees, and human beings all support the idea that the evolution of large brains is the result of selection pressure for succeeding in the manipulation of fellow individuals in a group. This idea, first proposed by A. Jolly and later by N. Humphries, has now been given the appropriate name of "Machiavellian intelligence" (Byrne and Whiten 1988).

As is evident, peck order and all the more advanced forms of dominance that I have mentioned briefly provide excellent examples of behavioral competition. The interaction among individuals for success in reproduction may depend upon being able to dominate and outmaneuver fellow individuals within a group. The ability to dominate and maneuver others is generally not directly determined by the genetic constitution of the individuals. In the case of insects, hens, and other animals with relatively small brains, the success depends on the state of the individual, that is, on the health and all the factors that affect health, such as age, nutritional state, and so forth. Those animals that are generally vigorous and healthy will be the victors. In the case of chimpanzees and other animals with elaborate dominance systems involving alliances, it is not just brawn that leads to success but cunning as well. Therefore, there will be a selection pressure for genes that lead to intelligence, but, despite this genetic background, the establishment of a hierarchy sequence itself, and all the maneuvering that goes with it, is solely a matter of behavior between the individuals.

Competition by Invention

It is often said—in fact, it is the traditional view—that human beings are capable of many mental activities that are totally denied to lesser animals. In recent years there have been a few hardy biologists, of which Donald Griffin (1992) is at the forefront and the most eloquent spokesman, who point out that we simply do not know what goes on in the mind of an animal. How can we be so certain that animals do not have some sort of self-consciousness, some sort of ability to rationalize, some ability to use symbols in communication, and the ability to solve relatively complex problems? The assumption that we are better at all these things compared to other animals is quite reasonable. Opinion diverges on the question of all-or-none: there is the ancient view, still held by many, that human beings are totally different and are endowed with mental characteristics that are completely absent in other animals. It is a legacy of the notion that we are specially created in the form of Adam and Eve, and in the image of God. In recent years, those who have thought about the problem find it very difficult to discover what goes on in the mind of an ape or a dog, and there is a legion of workers doing fascinating experiments to see if they can find a window into the animal brain. This has led to the opposing view: that there is a possibility that the only difference between *Homo sapiens* and its close relatives is one of degree, and it may be presumptuous to say that there are things that only we can do, and that we are in all ways unique.

The one place where one finds an impressive distinction between human and beast is in the sphere of inventions and the production of artifacts, although it seems to me that again this is a matter of degree. Primates are very clever in thinking of new ways of obtaining food, for example by reaching for bananas with sticks and boxes, and by removing sand from wheat kernels by dumping them in water and skimming the clean kernels off the surface; these tricks come very close to being inventions. However, in each case the amimals express their cleverness and create their innovations with tools they have at hand, with little or no further "improvements." A chimpanzee captures termites with a thin stick or plant stalk that it might prepare for efficient probing. In the case of human beings, no doubt early man did little more than this, and the great rise in the sophistication of our tools and their use arose through a cultural evolution, by the copying and the improvement of our predecessors' ideas.

There have been a number of authorities who have been interested in the evolution of human inventions from simple artifacts to complex machines. (For recent discussions, see Basalla 1988; Petroski 1992.) The concern has been partly descriptive and partly an attempt to understand the mechanism whereby the nature of the technical, human-produced objects change over time. It is primarily an interest in history, as well as cultural and socioeconomic factors, that lead to a particular succession of innovations. These are most interesting subjects that have been, and will no doubt continue to be, ably examined and discussed; but my approach and my argument as to the cause of the invention of artifacts and machines will be quite different.

I begin (as do all others, including Karl Marx—see Basalla 1988) with the fact that all inventions, from the axe and other simple tools to the modern complex machine, are the products of the human brain. To this foundation I now add the assumption that all inventions serve the same basic function: they are designed and built by human beings to give them a competitive advantage. The axe can more effectively kill for food or chop wood for fuel; the complex machine, be it an automobile or a microwave oven, is made to compete in making money, either for the manufacturer or for the user. The fact that we can transport ourselves or cook more rapidly is what makes these machines desirable, but the competition is driven by money. In our civilized world money represents food and fuel and all the necessities of life.

What I want to show here is that competition is the fundamental motivation for all inventions. Innovations provide a way of either directly competing with other human beings or indirectly competing by simply enhancing the chances of survival. Let me now develop this theme in greater detail.

All of the simplest and earliest implements were used to increase the efficiency of capturing food, providing fuel, making shelter, and fulfilling other similar basic needs. Later, with the advent of agriculture, tools for facilitating the growing of plants, such as plows, were invented. This in turn led to a need for more efficient modes of transportation, especially for moving produce and people. Each one of these advances in efficiency was in a very real sense due to competition. Survival and the betterment of one's condition is part of the struggle for existence, and here it is being carried out not by genes, but by memes—in this case rather special ones that are innovations, inventions. Early human beings were competing with nature and among themselves, and any mechanical device, be it a tool or a weapon, increased the chances in the competition. Furthermore, as in an arms race, if a tool was invented by any one individual, the neighbors quickly had to imitate it in order not to be at a serious competitive disadvantage. Therefore even the most primitive artifacts are the product of the human mind and were produced to improve the chances of their inventors excelling in the nongenetic selection that inevitably occurs.

As the history of humankind unfolded, tools became more elaborate, and eventually machines emerged. Again they are connected with the gathering of food: carts and boats for carrying (and fishing), bows and throw-sticks for hunting, buckets on wheels for irrigation, devices for raising water from wells, and so forth. The principle here is no different from that of simple tools: competition to survive more effectively. Each mechanical innovation or improvement is an advance in efficiency.

In the Western world these devices became more complex along with their increasing effectiveness. By the eighteenth century we had elaborate carts and carriages, printing presses, all sorts of clever water-driven machinery to raise water and drive power tools, and a vast multitude of other resourceful gadgets. As we look at Western civilization, we see a steady progression through the centuries of rapid advances in the design of mechanical devices, and if we examine Chinese technological history, as J. Needham (1965) has done in his monumental work, we see a parallel progression in the East.

In the nineteenth century a radical invention in the form of the engine changed our lives. First came the steam engine, followed by the electric motor and the internal combustion engine. These are devices that convert one form of energy (steam, electricity, gasoline) into another (mechanical energy). More than that, they are now so complex that they have their own internal dominance hierarchy, with a sequence of internal actions that has a beginning and an end. They may even have a master control (an on-off switch) and other features that clearly spell out an elaborate and orderly internal composition.

There is an interesting common property in living and mechanical machines: neither can ignore the restraints set by scaling laws. Engines, bridges, and all the creations of the engineer must take into account that weight rises as the cube of the linear dimensions, while strength and diffusion rise as the square. This means that in both cases—living and nonliving—there must be a strict adherence to these physical constraints. For living organisms, if these laws are ignored the individuals will be eliminated through natural selection; in machines such neglect means collapse, not only of the machine but very likely of the company that makes them as well.

The point I want to emphasize is that no matter how elaborate these machines may be, they are the product of the human mind—we have put our designs into them. They are our brain children, so whatever abilities they might have are our doing. The machine itself does not exist without our imagination and the transmission of our ideas. Furthermore, we have invented machines in our own image in the sense that they, like us, are involved in competition, which as we have seen has its beginning in the very origin of life. To say how remarkable it is to have machines that can convert energy or transmit signals, thereby imitating living organisms, is to miss the point. The machines are made by us—we gave them those properties, they are an extension of us.

Computers are the most recent and the most sophisticated of our machines. We have built them so cleverly that they can do things we cannot do with our brains; they have been designed specifically to make up for some of our deficiencies. The important point to remember, however, is that we built them, so that what they can do has been imparted to them by us. As a result it is not surprising that besides their great use in performing exceedingly rapid calculations to help the mathematician and others in many fields of science solve difficult and lengthy problems, they also have a great appeal in their ability to play games. It is not only chess, but all those video games for children. There is no more obvious manifestation of our own interest in competition than our love of games—from sports, to cards, to roulette, and now to computer games. We have programmed them so that they have become extensions of our competitive drive.

Two Interesting Footnotes to the Rule

Thus far the point has been made that if, at any level, there are a number of identical or similar units, they soon sort out so that some become dominant and others subservient, and that this is the result of competition. A number of workers, of which Deneubourg and Goss (1989) are a good example, have shown that a group of similar units can collectively produce order, and they do so not by electing a captain but by remaining totally democratic and leaderless. The examples they give involve the behavior of social insects, schools of fish, flocks of birds, and many other types of social aggregations. Their main argument is that while the behavioral capabilities of each individual in the group may be the same, each one can stimulate or influence its neighbors by what amounts to autocatalysis. The initial stimulus may come from an outside cue such as the distribution of food, as in the example of army ant foraging patterns, but the followers take their cues from the sister ants that surround them. Although the responses of each ant are only to its immediate neighbors, the end result will be a specific pattern of foraging of the entire colony, with an expanding fan-shaped front that is highly efficient in trapping prey. They propose the same kind of autocatalytic responses to account for the sudden turns and twists of flocks of birds and schools of fish. None of these patterns requires a dominant leader; they arise entirely from the autocatalytic behavior of essentially identical individuals. It is interesting to note that these patterns in social animals are explained by a simplified version of the appearance of pattern in developing embryos by reaction-diffusion mechanisms, where again autocatalysis plays a key role.

The second footnote has to do with those instances in which there is a disadvantage to dominance. A number of workers have observed, especially from behavioral studies on primates, that often the dominant individual has to pay a severe price to maintain its superior position. For instance, alpha females in a monkey group may end up with fewer offspring than ones lower on the social scale—the effort required to remain top monkey has taken its toll. The moral is that political ambition exacts a price, even in lower primates.

Neither of these two interesting points undermines the basic argument concerning the ubiquitous role of competition; however, they do serve to emphasize that competition is a subtle subject with numerous dimensions.

CHAPTER 3

Gene Accumulation and Gene Silencing

FOR MANY years I have been both puzzled and interested by the fact that often seemingly identical processes or traits are either entirely genetically fixed, that is, programmed by the genes, or they are somatic and their destiny is flexible and entirely determined by either external or internal signals.

I would like to begin by stating as simply and as clearly as possible what seems to me the essential phenomenon. This will be done as though there were no history, no background to the matter, and then I will briefly discuss that history. The reason for my doing it this way is that there is some disarray in the earlier literature that might obscure the main point I want to make.

Neo-Darwinism makes the simple assumption that the selection of individuals results in a change in the frequency of the alleles of certain genes in a population. At the same time, largely through the work of M. Kimura (1994), there is an appreciation of the fact that many gene mutations have no selective advantage and therefore must be considered selectively neutral. Their tendency to appear or disappear is entirely by chance and depends on mutation alone—selection plays no part. Needless to say, the evidence for both neutral genes and genes which are positively retained or eliminated by selection is overwhelming, and no one questions either of these well-established phenomena.

Consider now two possibilities. One is where a completely phenotypically controlled process that is not genetically inherited happens to occur repeatedly and consistently for generation after generation in an unvarying fashion. Eventually genes might appear by chance which will direct the same process that the organism was previously carrying out without any genetic control at all. I will call this process whereby genes fill a void created by the unchanging phenotypic activity *gene accumulation*. It could occur in a number of ways, such as the modification of existing genes or of new genes produced by gene duplication. To give an example, imagine a particular pattern of behavior that is not inherited but is continually repeated for many generations. Assume that new gene changes, which occur by chance, program that behavior, so that it now assumes a rigid, invariant pattern. This genetic fixation to produce an instinctive behavior pattern will occur entirely by the accumulation of neutral genes, that is, there will be no selection against them. This must be thought of as a stochastic process that occurs automatically if there is an endlessly repeating somatic event.

The second possibility goes in the reverse direction; one can have a process that is genetically fixed, and it will evolve into a flexible system that relies on signals that come from outside the genome. This I will call *gene silencing*. The genes themselves are unlikely to be eliminated, but rather their ability to act, to show an effect on the phenotype, simply has been turned off. This could happen by the influence of modifying genes, or the turning off of a promoter, or the alteration of the genes themselves. In all these cases the original genes would cease to function.

Again the easiest way of illustrating the phenomenon is to give an example from animal behavior. The whole evolution of learning would fit. For instance, there is obviously a great advantage for an animal's survival if it can recognize different prey, or different predators. In primitive animals all behavior presumably comes from fixed action patterns which produce a single, invariant behavior. An animal with the ability to learn has the opportunity of selecting prey or avoiding a particular predator using the best evasion technique for that predator. Somewhere along the course of this particular evolution, the genes that controlled the rigid ancestral action patterns must have been silenced—that is, selected out or blocked in some way—and replaced by genes which permit the more flexible learning. This would be an example of gene silencing by active selection, and as we shall see there are many other candidates for this phenomenon.

In the case of gene accumulation by the simple repetition of a nongenetic living process, neutral genes can creep in and direct the same function. These genes are neutral because their appearance is selected neither for nor against. At a later stage they may impart a selective advantage and subsequently be retained by positive selection, but that has nothing to do with how they arrived in the first place.

Gene silencing differs from gene accumulation not only in the direction of the genetic influence, but also in the fact that the silencing is under active selection; the removal of the expression of the genes is not a neutral process. It happens because there is a selective advantage to eliminating the genetic rigidity and shifting to a more flexible system—one which depends on external signals.

If one looks at the past history of gene accumulation in its various guises, it had its origins at the end of the last century. The topic has been admirably reviewed by G. G. Simpson

(1953), who shows that J. M. Baldwin, Lloyd Morgan, and H. F. Osborn all came up with the idea independently. They were motivated by their desire to bridge the theory of Lamarckian inheritance of acquired characters—which was an acceptable and a prevalent view at the time—with Darwinian natural selection. Their work was done before the discovery by Mendel became known, yet at the time it was clearly understood that there was a fixed, inherited variation which somehow had to be reconciled with the ephemeral nature of the effects of the use and disuse of organs on their immediate structure. The idea that a repetition of some somatic process over many generations could eventually become hardened into a truly inherited variation has been called the "Baldwin effect" by Simpson, a term in general use today.

In this century there are many who have considered the Baldwin effect. In particular I. I. Schmalhausen (1949) understood the problem with clarity, and C. H. Waddington (1957), with his combined interest in development and evolution, was prominent in stressing the importance of the principle, which he called "genetic assimilation." He did some interesting experiments with *Drosophila* on bithorax and crossveinless mutants. In the latter case he selected those few flies that lacked a cross vein in their wings after being given a heat shock. With repeated selection of these crossveinless flies, the percentage of crossveinlessness rapidly increased until ultimately, by the fourteenth generation, some of the flies emerged crossveinless without a heat shock. He interpreted this not as the appearance of new genes, but the result of selecting among the genetic variability already present in the genome (see Hall 1992 for an excellent review). Genetic assimilation differs from gene accumulation in that it can occur in a few generations, while gene accumulation would no doubt involve an exceedingly large number of generations.

The classic examples which he and others cite come from D'Arcy Thompson's *On Growth and Form* (1942). Here Thompson pointed out that abrasion causes the soles of our feet to grow thicker, as also occurs on the ventral callosities of an ostrich where the bird touches the ground when it sits. While these seem to be straightforward physiological responses to a mechanical stimulus, that is clearly not solely the case for these examples of skin thickening. In the human fetus the feet thicken before birth, as do the callosities of the unborn ostrich. These instances could be explained by genetic assimilation if one assumes an alteration in the expression of genes that were already present and responsible for the skin's response to abrasion. It could also be that there has been an accumulation of genes which act to thicken the skin, simply because there has been no selection against such neutral genes.

In this shifting back and forth from genetic rigidity to somatic flexibility, it is much easier to understand how genes can accumulate than to understand how one can shift to a more flexible system. One way of looking at the problem is to recognize that there will be some somatic variation even in genetically identical individuals. For instance, there are old experiments on the size of *Paramecia aurelia* which show that in a clone one can have a variation in length in which the smallest will be almost half the length of the largest, yet if one starts new clones from these extremes, their asexually produced progeny will show an identical range of sizes

(Ackert 1916; Jennings 1920). What seems to be inherited is the ability to vary somatically within certain limits, and I have called this kind of variation *range variation* (Bonner 1965). The basic point is that this variation may be just "noise" as Waddington (1957) suggested, or in some cases this noise could be captured to be put to work as a stepping-stone to flexible, somatic variation.

My first realization of such a possibility came in thinking about the development of cellular slime molds. It is easy to raise a population of cells from one spore (something that must occur some of the time in nature), and when one does this, as in *Paramecium*, there is a great variation in cell size within any one aggregate that forms an individual multicellular organism. The difference is that in this case it would appear that the somatic variation is put to good use. To explain how this is so, first let me briefly describe the development of *Dictyostelium discoideum*, a slime mold that has been much studied.

Cellular slime molds are soil amoebae. They feed as separate individuals on bacteria, and after they have finished the food supply they stream together to central collection points to form a multicellular individual of thousands of cells. This mass of amoebae moves as a unified "slug" toward light and is also oriented by heat gradients. After this period of migration, the anterior cells turn into stalk cells that keep piling onto the tip, while most of the posterior cells turn into spores. The spore mass is slowly lifted into the air as the stalk elongates; the final result is a small fruiting body, in the order of one or two millimeters high, in which a spherical spore mass is supported on a slender stalk made up of large, vacuolate, dead stalk cells (fig. 5).

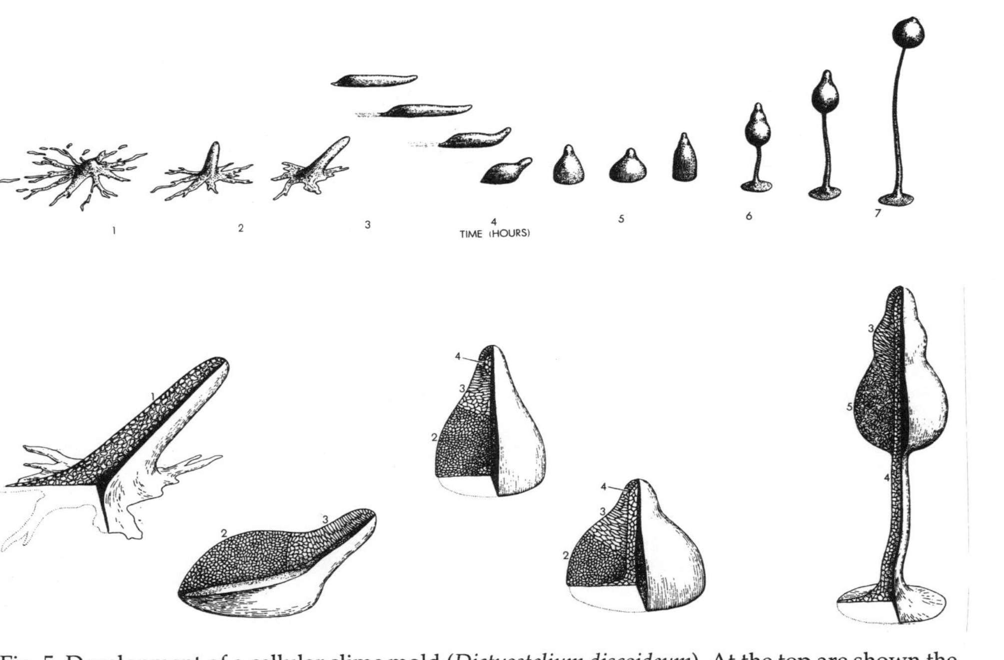

Fig. 5. Development of a cellular slime mold (*Dictyostelium discoideum*). At the top are shown the aggregation, migration, and culmination stages in an approximate timescale. At the bottom are cutaway diagrams to show the cellular structure at different stages: (1) undifferentiated cells at the end of aggregation; (2) prespore cells; (3) prestalk cells; (4) mature stalk cells; (5) mature spores. (Drawing by J. L. Howard from Bonner 1959, *Scientific American*.)

At first it would appear that just being at the front end of the slug ensured that a cell became a stalk cell, and a cell from the posterior portion became a spore. This was an early and obvious deduction from Raper's (1940) classic experiment in which he cut a slug into segments, and given enough time, each segment produced a normally proportioned fruiting body. However, in experiments I did some years ago (while on sabbatical leave in Waddington's laboratory), this turned out not to be the whole story: some of the cells sort out during aggregation and for a period thereafter, and some move to the anterior end and others to the posterior. In other words, some cells, already before aggregation, have either stalk or spore tendencies, and they seek out their position in the slug. Later it was shown by Leach et al. (1973) that if one mixed well-fed cells with ones grown on a minimum of food, the former would move to the posterior prespore region of the slug and the latter into the anterior prestalk region. This was followed by the discovery of McDonald and Durston (1984) that amoebae deprived of food at the beginning of their cell cycle ended up mainly in the prestalk region, while cells deprived late in their cycle became prespore cells. Clearly the differentiation of the fruiting body into stalk cells and spores of controlled proportions involves both within-cell signals (that is, signals that are direct gene products), and between-cell signals (which involve morphogens or signal substances, and their receptors will pass information to different parts of the cell mass to produce what Lewis Wolpert aptly called "positional information"). The latter is evident in slime molds because any cell that happens to be in the anterior end becomes a stalk cell, even if it had been a prespore cell. And the reciprocal is true for a prestalk cell that finds itself in the posterior prespore zone. On the other hand, within-cell signals dependent on the inner constitution of a cell will decide if it moves anteriorly or posteriorly in the slug. There is more to this story (for a review see Nanjundiah and Saran 1992); but the essence is that the cells may be genetically identical (if they are grown from a single spore), yet partly because of their somatic differences they manage to differentiate into two cell types. They have taken advantage of the range variation and put it to good use. Presumably if one could manage the impossible task of getting cells identical in all respects to aggregate, the cells would be incapable of differentiating into stalk cells and spores.

Asexuality

Before discussing behavior, where the same principle applies, I would like to examine some examples of gene accumulation and gene silencing taken from other basic biological processes. It is easiest to find cases which solely involve gene silencing. A case in point is the loss of sexuality in many animals and plants which then proceed to multiply totally by asexual means. This is especially frequent in animals and plants that have a normal choice between the two kinds of reproduction. For instance, in many freshwater green algae such as *Volvox,* sexual reproduction is something that occurs at the end of a vigorous growing season, and the result is a thickly coated, resistant zygote that is capable of surviving adverse conditions, in particular the rigors of winter. But in summer, when the growth conditions are ideal, the best strategy is to reproduce as rapidly as possible, that is, to reproduce asexually. It is easy to imagine that such an organism might end up in some environment where the seasons and the conditions in general were so constant that the stimulus to reproduce sexually is not called upon for long periods of time, and with such an absence of selection, the possibility that by some mutation or genetic change the ability to do so became lost.

This clearly must be the case for many fungi. There is a large group that is quaintly known as the "Fungi Imperfecti"; they are either Ascomycetes or Basidiomycetes that have lost their ability to reproduce sexually. Since these imperfect fungi make up a large and ancient group, one can only assume that their environment was such that they could cope with the loss of sexuality without becoming extinct. There are other plants and even animals that now rely wholly on asexual development, although none have done this with the success of fungi.

Development

A somewhat more interesting, albeit more speculative example may be found in the switch between mosaic and regulative development. It is more interesting because it involves both the accumulation and the silencing of genes. The suggestion that development should be examined in the light of gene accumulation was first made by S. Newman (1992, 1994); here I want to apply it to mosaicism and its lack in development.

In regulative development, if the embryo is divided into parts at an early stage, then each half will "regulate" and become a new entire whole by forming two miniature embryos. This regulation can be achieved only by signaling between cells. In mosaic development each cell has been programmed so that it cannot change its fate. If part of an embryo is excised, then the larva will miss that part. The signals are within cells and passed down through a succession of controlled cleavages, and for this reason it is also called a stereotypic cell lineage.

The details of this ancient embryology are too well known to be repeated here. Instead I want to examine a specific question which pertains to these two kinds of development. Why is it that some organisms, often only distantly related, appear to have a much greater emphasis on one of these two types of development than on the other? Mosaic and regulative development seem to be so totally different, yet whether the development of a particular group of organisms is mosaic or regulative at first glance seems quite arbitrary.

It is important to stress that no animal or plant has a development that consists solely of within-cell signals (except single-cell organisms!) nor one that consists solely of between-cell signals. Take, for example, the nematode *Caenorhabditis*. It has strict cell lineages in which the number of divisions, and the designation of which particular cell divides, seems to be entirely programmed in the genes—a clear example of development by within-cell signals. Yet now it is well known that for final differentiation there are signals between cells that play a crucial role, as for instance, between the anchor cell (AC) and the uterine precursor cell (VU). The AC cell controls the differentiation of the VU cell by inhibiting it from becoming an AC cell. The AC cell can be removed by a laser beam, and the cell that was to become the VU cell will turn into an AC cell. On the other hand, if a VU cell is killed, the presumptive AC cell will remain an AC cell. The between-cell signal goes from the AC cell to the VU cell. Already quite a bit is known about the signaling, which in this case involves two cells that are touching each other (see Greenwald 1990 for a review).

Conversely, organisms that have a highly regulative development, such as hydroids, also show evidence of mosaic development. Regulation is not only seen in the regeneration of the stem, as is well known in *Tubularia*, but it is also evident in early hydroid development, as is clear from G. Teissier's (1931) experiments on *Amphisbetia*. He showed that it was possible to separate the early cleaved cells as late as the sixteen-cell stage, and each isolated cell produced a perfect, miniature larva. He also cut early embryos into halves, and each fragment would form a perfect larva. Yet C. N. David and R. D. Campbell (1972), in their study of the interstitial cells of *Hydra*, found that they are stem cells and that the cells they produce, depending on the number of subsequent divisions, produce either nerve cells or nematocytes (to become the stinging cells, or nematocysts). In other words, hydroids show extreme regulative development alongside mosaicism by cell lineage. The fact that there is a mixture of between-cell and within-cell signals is further supported by David and Campbell's evidence that the cell lineage system has between-cell signals to regulate the proportion of nerve cells and nematocytes.

Let me now return to my original question: Why is not all development of one type or the other—Why are there such big differences among organisms? My hypothetical argument will be that the reason lies in natural selection and the relaxation of selection.

If the ability to regenerate or the ability to replace lost parts is of selective advantage, there will be strong selection for between-cell signals. On the other hand, if there is no such selection and development remains undisturbed for generations, it is quite possible that the details of development come under direct genetic control through the accumulation of new genes, that is, development relies mainly on within-cell signals which are the direct gene products. To switch into one direction or the other is easy to imagine, for both kinds of signaling invariably appear to be present in multicellular organisms.

For example, in hydroids it has long been argued that their embryos are subject to the battering of waves, and for survival a highly regulative embryo would have distinct advantages, which means between-cell signals would be favored. However, Leo Buss has pointed out to me that because of their small size, they live in a world of low Reynolds numbers, and therefore water is to them like viscous molasses, and wave action cannot be a danger. Perhaps there are other menaces, such as small prey that nibble at them; we do not know the answer. The wave or the predator argument could more easily be applied to the amazing ability to regenerate found in the larger hydroid polyps. One also wonders why mammalian embryos, at early stages, are regulative. Could it be that there is a need for flexibility before the stabilizing condition provided by implantation on the wall of the uterus? In all these instances one worries about speculating way beyond the facts.

As an example that has a much more obvious hypothetical explanation, the ascidians provide a perfect case of mosaic development. The fate of each cell in the embryo is laid out at a very early stage, and the cell lineage is remarkably rigid. Only at later stages in the development of the embryo does one find any significant between-cell signaling. Yet despite this extreme mosaicism it is well known that in clonal ascidians, which form large colonies of individuals connected by stolons, new colonies can bud from any portion of the stolon—they have complete powers of regulation. In one organism one finds the two extremes in different parts of the life cycle. Obviously regulation is required for clonal development but is unnecessary for early development. The latter may have become mosaic by default, that is, by the lack of selection pressure for regulation. Note that in this case, repeating the same process over and over again would not only fix the new genes, but would fix the cleavage pattern and the distribution of the substances responsible for differentiation in the different blastomeres. It would be a fixation at two levels: the rigid orders from the genome, and the rigid carrying out of those orders by the messenger RNAs and the proteins in the cytoplasm.

Higher plants provide another useful example. There is well-accepted evidence that all the cells of an angiosperm are potentially totipotent, and if a single cell is isolated and grown in tissue culture it can produce a whole new plant. At the same time, all the cells are imprisoned in hard cell walls and cannot move about. This means that although the cells are potentially regulative, due to the restrictions of their thick cellulose walls they are by their constrained circumstances forced into an inescapable mosaic state. Because of the immobility of the cells, and because of the large size of the plant, it is an absolute necessity that the growing plant rely on between-cell signals. It could respond in no other way to environmental stimuli which elicit such responses as phototropism and geotropism; they have to have some way of communicating between parts of an individual plant; otherwise, with their enforced mosaicism, they would produce identical plants in all particulars, regardless of the environment. The problem is solved by growth hormones and other kinds of chemical signals which influence the growth zones, or meristems, and all the other soft portions of the plant where new cell divisions and growth can occur. In this way plants are produced that can adapt to a particular place in the environment. Here the between-cell signals again are obviously advantageous and must be under positive selection. Let me add that this between-cell signaling also provides exceptional opportunities for regeneration, again a clear advantage and consequently encouraged by natural selection.

From these various examples it is evident that sometimes all, and sometimes one part, of a life cycle will have changes governed predominantly by either within-cell or between-cell signals, although there will always be some degree of mixture between the two. In the case of regulative development we see many examples where the between-cell signals have an obvious selective advantage. This is especially clear where there is restitution or regeneration after damage. There are other reasons as well, such as the requirements of a clonal organism or of a large, multicellular plant built with rigid cells that needs the between-cell signals simply to grow in a consistent manner.

All in all there is a fair case for the selective advantages in many instances of regulative development, although there are some instances where it is unknown what the advantages might be. This is not so for mosaic development; it is hard to imagine what might be the advantage of an increase in rigidity. Perhaps in most instances there is none, and mosaic development that relies primarily on within-cell signals is a case of gene accumulation. Imagine a moderately regulative development that occurs without any modification over many generations. If there are mutations that arise by chance and specify the same developmental steps, those genes will accumulate and give within-cell signals that supplant the more flexible between-cell signaling that reigned previously. This will work provided there is no selection for flexibility; the moment there is, the new, rigid, gene-controlled steps will be selected out by gene silencing, and the flexibility of between-cell signaling is restored. If this view is correct, regulative development is under active selection pressure, while mosaic development occurs by gene accumulation due to the constancy of between-cell development over long periods of time. There could be exceptions to this proposed hypothesis. For instance, it is conceivable that there may be some circumstances where the mosaic, within-cell signaling is selectively advantageous, in which case genes that control those signals will be actively retained, and the genes that control the between-cell signals are silenced. In any event, it is reasonable to assume that these different patterns of development are a direct product of Darwinian natural selection.

Aging

Gene accumulation can be seen in the process of aging, and it is a particularly helpful example because it adds another dimension. It was primarily Medawar (1952) who established two closely related ideas that are pertinent here. (Haldane and subsequently many others have contributed to the matter; for a review see Rose 1991.) As an individual becomes older, the number of potential offspring it could produce automatically declines, and therefore any deleterious mutant gene that makes its appearance specifically during old age is less likely to be selected out. Because of the diminished reproduction, or perhaps even its absence, the gene will be neutral or approach neutrality. Such a case would fit perfectly into what I have been calling "gene accumulation." The other idea is that there could be a mutation that is strongly favored in the early reproductive phase of the organism which has a pleiotropic deleterious effect later in life when it essentially no longer reproduces. Here we have something added to neutral gene accumulation: the neutral effect is produced by positive selection because both results are produced by the same gene, which has two different effects at two different times in the life cycle. It should be added that, as always, these changes in the phenotype could be due to modifier genes, and the actual genetic basis of the change could be complex.

Under such a system it is easy to see that senescence could be the result of a wide variety of deleterious mutations that have crept in by one of these two ways. Either way—being fixed by pleiotropy or being neutral because they affect the phenotype in a period after active reproduction—deleterious genes that give rise to a great number of disorders and failures that are associated with old age can accumulate. As numerous authors have pointed out, by looking at aging from this evolutionary point of view, one would expect that the physiology of aging does not involve one process but many. The randomness of the mutations that appear late in life assures that the body can break down in a wide variety of ways.

Atrophied Organs and Atavisms

For completeness I must mention that, as has been universally recognized, gene silencing is illustrated by the common presence of atrophied organs in animals. The appendix and the coccyx of human beings are classic examples. They conveniently illustrate another pertinent point. Sometimes the reduced organ will revert to its ancestral condition: for instance, occasionally a human baby is born with a tail rather than a hidden coccyx. Such atavisms show that the ancient genes have not necessarily been eliminated but simply lie dormant. It could be that there has been a mutant modification that has weakened a suppressor gene, or the gene itself might have undergone some modification that permitted it to become dominant and capable of expressing itself.

Silencing is a very general principle that can be seen throughout the animal and plant world. Even bacteria exhibit the phenomenon in a striking way. It is not just that it is possible to turn on certain inducible genes by producing the appropriate substrate, but it has been shown that there are whole hosts of genes that exist within the genome but are totally silent. The only way they can be reactivated is by mutation. At first glance it may seem wasteful to keep this bank of unused genes taking up space on the chromosome, but in the long run it may be a most effective strategy for survival. Bacteria have very rapid successive generations, and for that reason their effective mutation rate is high. This means that should they land in a region where there is a food they cannot use but their ancestors could, there is always the possibility that the gene which has been lying dormant might mutate to its active form. If one individual goes through this mutation, its progeny will flourish, while all its siblings will perish. For these reasons it is not surprising that, for long-term survival, bacteria have a large number of such hidden genes in their genome.

Behavior

The case of switching from a flexible, somatic state to one that is rigidly controlled by genes is certainly best illustrated in animal behavior. It is the original basis for the "Baldwin effect," and because today we understand the role of genes, we can make a much stronger case. Let me begin by briefly mentioning some well-known and obvious examples of shuttling back and forth between genetically fixed behavior and flexible learned behavior.

An ideal case is that of birdsong, which has examples of both behaviors, and there are quite reasonable explanations how each might be selectively advantageous. In many songbirds the offspring learn much of their song from hearing their parents sing, and apparently this is advantageous for reasons related to recognition. The young can not only recognize their parents, but also their whole group, for this kind of song learning leads to the formation of dialects characteristic of different groups. It is also used for mate recognition when pairing occurs in the spring. Furthermore, males use the great flexibility of the song for staking out territories and attracting mates; song is used by the males as advertising in sexual selection in the same way some birds use gaudy plumage to attract mates.

Another example would be the recognition of predators by birds. It has been well established by E. Curio (1978) in some famous experiments that the common practice of mobbing a predator to chase it away, or at least to alert other members of its kind of the danger, can be passed on from one individual to another by imitation. The obvious selective advantage of such flexible learning is that if a new predator enters the area, as soon as one bird discovers its evil intentions, the information can very quickly be spread to the rest of its kind. The opposite case is illustrated by some classic, earlier experiments of Konrad Lorenz, who showed that newborn goslings will rush under the bushes if the silhouette of a hawk should pass overhead. Obviously, in this vulnerable period of their early life, there is no way they could learn the danger of hawks, therefore selection for the response to be built into the genes would be strong.

Clearly there are many cases where a behavior that was initially flexible has become, conceivably by gene accumulation, innate and rigid; and similarly there are many cases where a genetically determined behavior has, conceivably by gene silencing, become flexible. And it is quite obvious that natural selection has been the driving force for the pushes in both directions, depending upon where the advantage lies. I shall return to this matter later, when I will compare it to gene accumulation and silencing in development. But first I want to examine a different way in which genes that affect behavior can infiltrate and accumulate.

In social insects the division of labor found within the insect colony is somatic and under the influence of external factors such as the supply of food or the effect of pheromones during the period of growth of an individual. To take an example from ants, if a worker is very small (due to a minimal diet as a larva), it will behave as a nurse and take care of the young larvae as well as perform other household duties. On the other hand, if an individual larva is fed on a rich and copious diet, then it will become large and behave as a soldier, guarding the nest. Workers of intermediate size will perform other tasks, such as the all-important one of foraging for food and bringing it back to the nest. It has been appreciated for a long time that the different sizes of individuals in the ant colony are not reflected in the genes—all the differences in the behavior are somatic and the result of how much they were fed while growing. In many species of ants the young queen starting a new nest, while still alone and before the emergence of her first brood, can and does perform the entire repertoire of behaviors of all the different workers she will eventually produce. If one experimentally removes one caste from a colony, the other workers nearest in size to the missing ones will immediately take over the chores of that group so that the colony can continue to exist without missing a beat. There is nothing to indicate that the different labors of the workers are in any way genetically determined, or even influenced by the genes.

Recently there has been an accumulation of evidence that this pure somatic story has some very interesting exceptions. The most dramatic discovery was that of R. E. Page, Jr., and his associates (for a review, see Page and Robinson 1991) that different genetic strains of honey bees differ in the behavior of their workers. Normally a queen will be fertilized by anywhere from seven to seventeen males, and she will retain this great mixture of potential paternity in her spermathecal sac. The inevitable result will be an impressive mixture in the genetic composition of the workers in the hive. If, in the laboratory a number of queens were each separately inseminated by a different male, their respective offspring differed significantly in their behavior. For example, some emphasized guarding activities, some foraged more intensively for pollen than for nectar, and some did the reverse. These differences are over and above the changes in tasks all workers go through with age—they clearly are genetic differences determined by a particular male. This means that in order for a colony to have a complete mixture of workers that performs all the necessary tasks, it is essential that a queen mate with numerous males so that she will be able to produce a well-balanced brood of workers.

It seems to me the easiest way of understanding this phenomenon is to think of it as an instance of gene accumulation. The infiltration of these task-favoring genes is absolutely bound to multiple mating: it could only occur if there is a mixture of task genotypes in the colony. This immediately raises the question of which came first. It could be either that task genes followed the appearance of polyandry, or that the two arose together. It seems hard to see how the task genes could have survived in queens that are fertilized by only one male. In other words, honeybees provide a good example of gene accumulation, albeit an unusual one.

Another most interesting way in which genes change in social insects can be seen in the division of labor between a queen and the workers. In primitive wasps there is no physical difference among the queens that group together to form a colony, but soon one emerges as more aggressive than the others; she becomes the functional queen through behavioral dominance. For instance, if any of the other females lays an egg in one of the cells, the dominant queen will eat it and replace it with her own egg. It is only in larger and more complex wasp colonies that the queen is larger than the other female workers. This size dimorphism, as well as worker sterility, seems to have appeared secondarily—the behavioral difference has evolved into a morphological difference.

Let us now examine this interesting case from a hypothetical genetic point of view. Initially there must be genes that favor aggression, but later in evolution those genes were replaced by genes which permitted the production of morphological differences. Perhaps they were genes that, through pheromone production, allowed the dominant individual to produce two different kinds of cells for rearing the young: one kind for the sterile workers, and a larger kind for producing reproductives, which come later in the season. Note that in all cases that involve varied nutrition of the young, a strong element of behavior is still involved—aggression has been replaced by controlling the amount of food for the next generation of workers.

If one were to look for a case where evolution has been in the reverse direction, from morphology to behavior, one could cite the transition from birds of paradise to bower birds. As Thomas Gilliard (1963) has pointed out, bower birds evolved from birds of paradise, and the former seem to have progressively replaced the spectacular plumage of the latter with equally spectacular behavior. There has been no diminution in the vigor of the sexual selection—it is simply a case of changing the means by which the males show off. The building of elaborate bowers with their extraordinarily colorful decorations, all put in place by the relatively drab male, are certainly a good behavioral equivalent of the plumage color and morphology of the birds of paradise. Again in this case there is no gene accumulation or silencing in any special sense; it is simply a case of gene change that would go with normal evolutionary changes.

These two examples serve a useful purpose for my argument. I have started off with the point that there are dramatic instances where genes have either been added through the repetition of some somatic process such as a behavior, or suppressed with the result that there can be a liberation of somatic processes to produce flexible responses. In the two examples I have just given there has been a shift from behavior to morphology, and the reverse. They involve a combination of genetic and somatic control, but whenever they are in a state in which behavior is predominant, that state is the more flexible condition. Behavior is the supreme system for flexibility. Therefore it might be best to consider these two examples as shifts between more and less flexibility. It is not the same thing as genetic rigidity and somatic flexibility, but it provides an interesting parallel. The examples are useful in that they show that the gene accumulation–silencing dichotomy is only one way of producing a rigidity–flexibility dichotomy.

Conclusion

I have made the point that there is, during the course of evolution, a shuttling back and forth between two apparently quite different ways in which living organisms are controlled. One involves a genetic rigidity that comes from gene accumulation, and the other involves a somatic flexibility that comes from the shedding of the strict genetic control by silencing the genes involved. I want to conclude by looking at how the two are related at a deeper level, and how it is possible to shift from one to the other. I have tried to show that this shift must have happened frequently during the course of evolution.

It has been appreciated for some time (and elegantly laid out by Gould and Marler 1987) that the ability of an animal to learn different things is not uniform: some kinds of information are much more easily learned than others. For instance, rats can learn to avoid foods that make them ill by identifying the smell, but they have difficulty avoiding such foods if the only cues they receive are the color or the shape of the food. In this case, learning by smell is obviously going to be more useful and advantageous in the wild than seeing the food, especially considering rats are nocturnal animals. Their learning capabilities will be those that are tailored to learn the functions that are most important to them for survival. Natural selection not only influences the best physiological way to detect the suitability of food, but also selects for the ease of learning that way. To give another example, many songbirds can only learn songs that resemble their natural song. In these instances there is evidence of an inherited, innate rudimentary basic song, and the learning involved in imitating other birds of the same species is the refinement of that song. Gould and Marler suggest this is perhaps the same as Noam Chomsky's argument that human beings are born with an innate basic grammar, and that learning language is simply building on that given, genetically determined structure. So again here we have an instance where flexible learning is built on a rigid, genetic program.

Because we do not understand how genes control behavior, it is difficult to visualize the mechanics of how selection for learning a particular set of cues can occur, but clearly it is something that happens. Somehow the ability to learn, and specifically to learn some things better than others, is a property of the brain that not only has a genetic base (or natural selection would not be possible), but one that can be influenced by selection so that when greater flexibility is advantageous, the ability to learn certain things can be increased. By putting the matter this way we can see that a shift to and fro from gene accumulation to gene suppression is not a wide gulf. Always present is the background of a somatic response, in this case the capability of learning, and the question to what degree it is encouraged or suppressed can be governed by selection. If a flexible system does not need its flexibility and does the same thing over and over again, then genes will creep in to make the system rigid. If a rigid system could use some flexibility for a selective advantage, then those genes that are capable of controlling the ability to learn will be encouraged, and in particular to learn those things that are useful to the individual and will increase its reproductive success.

Let me now apply these same arguments to development—specifically to the shifts back and forth between mosaic and regulative development which were discussed previously.

In the case of regulative development, where any part of the embryo can produce a miniature, whole embryo, it is clear that each part has all the information needed to make a complete, new individual as well as the flexibility to carry out those genetic orders under new circumstances. As I said earlier, it is easy to imagine that under some circumstances there might be strong selection pressure for such regulative flexibility, and as a result all necessary innovations such as the between-cell signaling system would have been encouraged. This would involve the cultivation of the response system, which would be attuned to respond to specific between-cell signals—or in some cases to environmental signals to coordinate the developmental timing with the season—and other essential environmental cues for survival. Therefore, in this case there is an interdependence between the signals and the responses which has been encouraged by selection to evolve in a way that allows multiple signals and multiple responses to produce regulative development with all its flexibility. Remember that the differences in the various parts of the embryo are entirely somatic, and therefore all the nuclei of all the cells contain the same genetic information. So even though differentiation is somatic, it can be established by virtue of the complex interaction of the between-cell signals. Particularly important is the response system to those signals, for there can be responses to both between-cell signals and the immediate environment of the embryo. Thus development flexibility makes it possible to keep order within the embryo under altered circumstances, and by responding to the environment it can have greater assurance of ultimate reproductive success.

Flexibility in development is, in one major respect, analogous to what we saw in the case of behavior. Flexible behavior is even more obviously the direct product of natural selection, for under many circumstances such behavior has clear advantages. What has been selected are multiple responses to different signals to cope with a varied and changing environment. However, behavior goes one big step further—it includes learning, something that has no counterpart in development. Learning which external signals are favorable and which are dangerous can be an extraordinarily rapid and efficient way to immediately choose the appropriate response. As we saw, the efficiency has even been increased by facilitating the learning of certain kinds of signals, so that not only has there been a selection pressure for learning, but a selection for making it easy to learn certain important categories of information.

Even though in general there is a constant selection pressure for an increase in flexibility, there will be some instances where the flexibility is no longer needed to survive or reproduce successfully. Should that be the case, then the same somatic process will be repeated generation after generation, allowing an infiltration of genes that will give within-cell signals that act in the same way and thus produce a rigidity by this gene accumulation. If the environment changes and opens the way for renewed selection for the flexibility, selection will opt for the silencing of those genes.

CHAPTER 4

Why Does Labor Become Divided?

ONE OF the most general and universal rules for living organisms is that they possess a division of labor. This may be found at all levels of organization: within the organelles of a cell, within groups of cells in the guise of differentiation, within groups of individuals in an animal society, and in our culturally determined human societies. Here I want to ask: Why is there such a general principle—what is its underlying cause?

There is an obvious correlation between size, complexity, and a division of labor. As entities become larger, they inevitably have more parts and are therefore more complex; this plethora of complicated parts does not, in living organisms, remain chaotic but becomes orderly by dividing the labor. Elsewhere I have argued that over the grand course of evolution there has been clear evidence for the selection of an increase in the upper size limit of living organisms, and that along the way there has been constant selection for both an increase and a decrease in size; these size changes are accompanied with changes in the degree to which their labor is divided (Bonner 1988).

The key question is, what produces the division of labor? Is it something that arises automatically in any complex mix of entities, or is it the result of the culling of numerous variants by natural selection? I do not subscribe to a rigid view of the former—that is, to the structuralist argument that there is an innate property of matter and therefore a division of labor appears to be an invariant characteristic of organisms. My difficulty with such a point of view is that it does not explain anything. It simply states that such a property exists. What that property is remains an abstract mystery, and therefore provides me with no satisfaction. (However, as I will discuss further on, there are some aspects of the division of labor that can be explained by the properties of similar units that exist in a group.) Much more appealing to me is the argument that the main cause of the appearance of division of labor in the evolution of living organisms is a natural selection for efficiency. Invariably a division of labor means an increase in efficiency, which in turn will automatically mean an increase in reproductive success.

Efficiency inevitably includes respecting physical laws as well as biological processes. This obvious point can be illustrated in a very simple way. If the size of an animal increases during the course of evolution, then those functions that involve surfaces (L^2) such as gas exchange and food assimilation must, in order to remain efficient, compensate to keep up for the increase in body size (L^3). This is the reason we find an increase in lung or gut surface in larger animals—and an elaborate circulatory system. Strength also varies as L^2, with the result that big animals have disproportionately thick limbs, and large trees have disproportionately large trunks.

Reproductive success cannot exist in a living structure that is physically impossible, or even inefficient in any way, so the surface-volume problems will always be dealt with and only structurally sound organisms will be the result. However, to make it clear that the organisms and their internal division of labor are not determined by, but only guided by, sound engineering principles, one need only look at the great many different ways in which the surface-volume constraints have been dealt with. Consider, for instance, the variety of methods that have been devised for bringing oxygen to the internal tissues. In air-breathing animals there is the tracheal system of insects and the lung and circulatory systems of mammals and birds, to give one of innumerable possible examples. In each case natural selection promotes a different way of achieving the same end imposed by the physical necessities. There is a massive degree of convergent evolution in the ways of dividing labor for solving not only the gas-exchange problem with increase in size, but for all the other surface-volume and strength-weight problems.

What has been said so far will seem obvious; now I would like to probe somewhat more deeply to see if one can find more direct evidence that natural selection produces a division of labor. Is it possible to catch evolution in the act, so to speak? An effective way of doing this would be to find instances where we can see the beginning of labor being divided and speculate on the advantages and disadvantages of the first steps. By reconstructing the initiation of the phenomenon one might be able to see more clearly how selection can play the crucial role. There are good examples at two levels: one is during the early stages of the evolution of simple, multicellular organisms, and the other is found during the formation of the most primitive insect societies.

The Volvocales

The *Volvox* group of freshwater green algae provides an ideal example of organisms that came into being by the adhesion of cell-division products, because living today one finds species of all different sizes—from four cells up to *Volvox* itself, which has thousands of cells. Depending on the species, colonies will span a size range from about 2,000 to 60,000 cells, the larger colonies being clearly visible to the naked eye. Since they exist at the present moment, it is safe to say that the ecological niches for those size levels also presently exist, but it is very unlikely that all the smaller living forms are the actual ancestral transitional types, although perhaps they are comparable. From recent molecular phylogenetic studies of D. Kirk and others (Kirk et al. 1991; Schmitt et al. 1992; Kirk, pers. comm.) it is clear that the forms that live today do not represent a linear phylogenetic sequence in which size and complexity have increased in incremental steps. Rather, they show that the genetic differences among various members of the volvocales are small, and with few gene changes they could shift from one type to another easily. This reinforces the idea that the niches today are the driving force for the existing gamut of morphologies. This does not preclude the possibility of a stepwise increase of complexity in their early evolution. There is reason to believe that the volvox line is ancient, for Kazmierczak (1981) has found fossils of *Volvox*-like organisms in Precambrian rocks.

The smallest existing members of the volvocales are some species of the genus *Gonium*, which have colonies of as few as four cells, while other species have 8, 16, and even 32 (fig. 6). The cells are held in a gelatinous matrix, and all their flagella point in the same direction so that in larger colonies they form a shieldlike plate. To give some other examples, colonies of *Pandorina* form small spheres of 8 to 32 flagellated cells; *Eudorina* colonies are somewhat elliptical in shape and

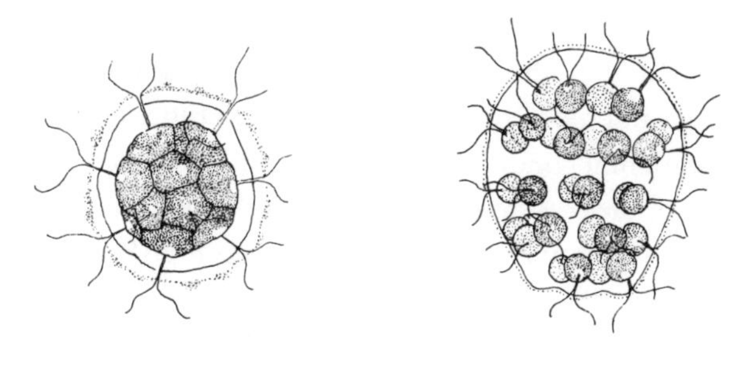

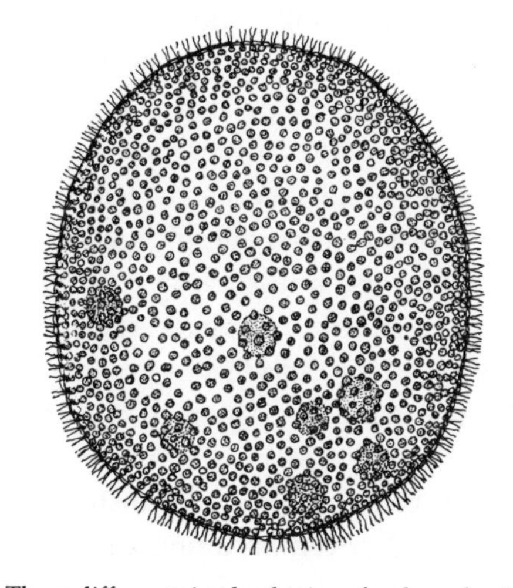

Fig. 6. Three different-sized colonies of volvocales. Upper left: *Pandorina*; upper right: *Eudorina*; bottom: *Volvox*. (From Bonner 1974)

are slightly larger, from 64 to 128 cells; and there are numerous other genera. However, none is so large as *Volvox*.

All the species have both a sexual and an asexual reproduction. In the summer season, when conditions are fairly constant and favorable for growth, a rapid sequence of asexual cycles is the norm. Sexuality appears at the end of this growing season, and the zygote produces a resistant cyst. It is guessed that this is a case of a familiar phenomenon where sexual reproduction occurs only in times of uncertain environmental conditions. By producing variable offspring, the sexual process hedges the bets for the new colonies germinating in the warm weather of the spring (Bonner 1958; Bell 1982).

A colony of *Volvox* begins by the cleavage of a large single cell (called a gonidium) and the daughter cells of each division remain attached laterally, with the result that they will produce a small, hollow sphere of cells within a vesicle inside the mother colony. The polarity of the cells seems to be inflexible, with the result that the flagellar ends of all the cells point inward inside the newly formed blastula-like sphere. After cleavage, this small sphere turns inside out by a process called inversion (Pocock 1933; Kelland 1977; Viamontes et al. 1979) ; the young sphere is now motile with the flagella on the outside, and when it eventually breaks out of the disintegrating, aged parent, it is a free-swimming colony.

In the asexual growth of a colony there seem to be two principal things that govern its ultimate size and form. One is the gelatinous matrix that holds the cells together, and the other is the number of cell divisions. Both of these can be linked directly to the action of genes. In the case of the matrix, upon instructions from the genes the enzymes are produced, causing the synthesis of the extracellular adhesive material. Our knowledge of the biochemistry and molecular biology of cell division is rapidly increasing, and there are many genes that control the cell cycle as well as substances which accumulate and fall with each cycle. It is easy to imagine the idea (and there is much evidence to support it) that there can be genetic control that specifies a particular number of divisions; dividing cells can have a counting mechanism.

It was shown many years ago by Bock (1926) that in all the volvocales, with the exception of *Volvox* itself, any cell in the colony is capable of generating a new colony. However, in the larger *Volvox* this is not the case; there is a division of labor. Only certain large cells, the gonidia, are capable of cleaving and producing a daughter colony. The large multitude of vegetative cells have lost this ability.

More recent work has shed some very interesting light on this old story of Bock. The first step was the discovery of Richard Starr (see his review of 1970) of a mutant of a species of *Volvox* in which all the cells of the sphere were capable of cleaving and producing daughter colonies. It appeared to be a reversion to the state of the more primitive volvocales. It has now been shown by Kirk (1988) that at least three loci are involved in this transition to *Volvox*. These genes control a separate lineage of cells that form the gonidia by stopping cleavage sooner than in the lineage of the vegetative cells.

It is interesting, as A. G. Desnitski (1992) has pointed out, that in some species the gonidia grow and become large, and when they develop into a daughter colony there is a rapid sequence of divisions producing many small cells in the new colony. In other species, which he considers to be a later evolutionary development, the gonidia grow continuously between divisions as they form a new colony, and therefore the time intervals between the divisions are greatly extended.

Order among a group of cells requires some kind of signaling so that the cells can function in particular roles in the colony. In the *Volvox* example I have just given, the signals which decide if a cell is to become reproductive (a gonidium) or vegetative are generated from *within* the cells. Also, whether a gonidium grows first and then rapidly divides, or whether it is the type that grows between divisions, is similarly determined by signals from within. These within-cell signals may come from direct gene messages, or they may depend on the cell reaching a critical size to evoke those signals. These within-cell signals are passed down in a cell lineage. Presently signals *between* cells will be considered where they are most prevalent—in aggregating organisms. It should be noted that *Volvox* also has between-cell signals: in the sexual phase, if one colony begins its sexual cycle it will give off a substance that induces other colonies to do the same, thereby encouraging the synchrony of egg and sperm production. All this can be summarized by saying that organization in primitive multicellular organisms (as well as more advanced complex ones) is achieved by a combination of within-cell and between-cell signaling.

In considering *Volvox* one might well ask: Why has a division of labor appeared? Why would it not be advantageous for all the cells to remain capable of division, and therefore reproduction? It is quite obvious that if a large colony has a selective advantage (perhaps to avoid being eaten by filter feeders, as Bell 1985 suggests), then size can be increased by increasing the number of flagellated cells on the surface of the *Volvox* sphere. Even if the cells are smaller, the colony can remain large due to the great number of cells and the gelatinous matrix that holds them together. Furthermore, it will swim more rapidly with a larger number of flagellar paddles protruding from it. It would also seem evident that the easiest way to achieve this size increase with the minimum reconstruction and reorganization would be to allow some cells to divide more frequently and as a result become small vegetative cells which, as a result, have lost their power of reproduction. At the same time a few controlled cells would remain larger by dividing less frequently and therefore remain capable of becoming the reproductive gonidia. By this clever plan size increase can occur with a minimum cost. It is true that there is a loss of cells capable of reproduction, but perhaps the remaining gonidia (which are usually between eight and sixteen in number) are quite sufficient to keep the population size flourishing under ideal growth conditions. Remember that reproductive success is not just the result of a large number of propagules, but also the survival of the individuals. If Bell's suggestion is correct, then the advantage of size increase to escape being eaten by filter feeders must be balanced against the decline in the reproductive rate due to the restriction of reproductive cells solely to gonidia. Finally, note that the genetic cost of this major evolutionary step is also small, for very few gene changes that involve within-cell signaling are sufficient to produce this division of labor. The possibility that these changes occurred by means of natural selection is self-evident.

Social Amoebae and Social Insects

Let us now turn to my second set of examples of primitive division of labor, this time in social organisms. Here two quite different kinds of beginning societies will be considered and compared: social amoebae and social insects.

The cellular slime molds become social by aggregation. As in the volvocales, there is a whole series of species, some of which are elaborate and have a clear division of labor while others are simple by comparison and are made up of one cell type only. As was the case for the volvocales, one cannot assume that this is an actual phylogenetic series (and there is strong evidence that it cannot be so, as we shall see), but the fact that there is a series of forms of increasing complexity living today is good evidence that there are ecological niches for each step in the progression.

Cellular slime molds are an extremely common and successful group of soil amoebae. E. O. Wilson (1990) has pointed out how abundant social insects are, and gives dramatic figures showing that ants, for instance, which are all social, make up a large portion of the animal biomass in a tropical forest. I have never attempted similar calculations, but since social amoebae are found in all soils and are one of the main consumers of soil bacteria, I expect they greatly exceed ants and other animals in all habitats in tonnage.

Because slime molds are hidden in soil they are not so well known, and it is only recently that we have acquired a detailed knowledge of their life history and their ecology, all admirably reviewed by K. B. Raper (1984) in his magisterial book. There are two major groups of cellular slime molds: the acrasids and the dictyostelids. They clearly have separate origins, for they differ conspicuously in the structure of their amoebae (Olive 1975). Since I will use examples of both to illustrate a single evolutionary sequence, my illustration obviously cannot be the real phylogenetic sequence; but hopefully it reflects the actual early steps, some of which have presumably become extinct.

In the acrasids the aggregation of the amoebae seems to be a simple accumulation of cells into a mound which forms a rudimentary fruiting body. Unfortunately, nothing is known of the mechanism of acrasid aggregation, although many of the details of dictyostelid aggregation have been revealed in the last fifty years. An attractant chemical (an acrasin) guides the amoebae into central collection points, producing a very dramatic inward streaming, and the resulting mound of cells forms a fruiting body made up of dead stalk cells within a delicate cellulose cylinder, and of terminal masses of amoebae that have become encapsulated into spores (see fig. 5). This is a clear division of labor, and the process of formation of the elegant fruiting body is an organized and controlled process. I should add that the cycles that I have briefly outlined are asexual. The dictyostelids also have a sexual cycle where separate amoebae of the opposite mating types fuse and form a giant zygote cell that encysts and produces new generations of genetically recombined amoebae upon germination. We will not be concerned with the sexual cycle here, but only with the asexual cycle in which we find the multicellular stage with all its organization and division of labor.

One way of looking at how complex dictyostelids and their division of labor evolved was inspired by my collaboration with R. Gadagkar, who is interested in the evolution of social wasps from solitary forms to ones with large colonies and a division of labor (Gadagkar and Bonner 1994). His work and that of many others has shown that there are a series of steps that can account for the evolution of a social existence of wasps, each one of which can be interpreted to be the result of natural selection. Briefly, the argument is that groups of females may find it advantageous to nest together, in that by doing so the number of offspring per female increases due to the communal care and the communal defense of the young. Within the group of females a dominance is achieved by one of the females becoming the queen, with the result that through her aggression she produces all, or a majority, of the eggs. R. D. Alexander and M. J. West-Eberhard and others (see Gadagkar and Bonner 1994 for references) have considered that this queen dominance has led to the manipulation of the other females, causing them to become workers and uninvolved in reproduction. If by chance some of these workers should become sterile, they will not be selected against, for they are already not reproducing in their subordination to the queen. So the division of labor between reproductives and sterile workers goes through a sequence: first a behavior-dominance system, then a morphological differentiation when the queen becomes noticeably larger. Note that all of these steps involve reproductive success and are therefore very much the products of natural selection.

Switching to slime molds, the first question one must ask is, what are the advantages for single, isolated amoebae to come together by aggregation? The most likely selective advantage of this cell grouping in the soil has to do with dispersal. Slime molds feed on soil bacteria that have a patchy distribution in the soil and in the humus. If one amoeba reaches a food patch, it will multiply rapidly; but when the patch is consumed it is necessary to disperse to find new patches. There is evidence that spores may be carried by various soil invertebrates, worms, insects, mites, and various other creatures, either by the cells adhering to their surface or by eating them and dispersing them by leaving droppings in distant places. If the propagules are bunched in a group, then the traveling invertebrate will be able to pick up a large number of propagules, thereby increasing the chances that one (which is all that is needed) will be left in a fresh patch of food. There are also solitary amoebae in the soil, and they obviously exist in a different niche, for they apparently coexist with the more social amoebae.

Let us first consider dominance: is there anything equivalent in slime molds to a queen wasp? There is no information in the acrasids, where no experimental work has been done, but it is well known among the dictyostelids. In *Dictyostelium* there is a small group of cells (or possibly a single cell—it is simply not known) which put out a chemical (acrasin) that attracts other amoebae by chemotaxis. This chemical is emitted in periodic pulses so that these initial cells are not only the first to emit the chemoattractant, but by emitting the chemical signal in regular pulses the central cells dominate the other cells. The reason is that there is a relay system so that when one cell emits the acrasin, it stimulates neighboring cells to emit acrasin in turn. In this way the central cells become the signaling command post, and the other cells can only pass on the signal.

So right from the beginning of aggregation we see the appearance of dominance, although it must be clear, as in the primitive wasps, that this dominance is not a genetic difference, but a somatic, physiological one. In the slime molds all the cells can be genetically identical, yet at the onset of aggregation they will form a peck order. What is genetically selected in this order is not which cell (or wasp) becomes queen, but the ability of the cells to differ in such a way that some cells become dominant over others.

This point is especially well illustrated in another genus of dictyostelids, namely *Polysphondylium* (see fig. 9). There, B. M. Shaffer (1961, 1963) made the most interesting discovery: that it has one clearly identifiable cell which is different from the others and is responsible for the initiation of aggregation. What he called a "founder" cell rounds up and becomes a focal point for other cells; they stream in towards it (fig. 7). He did many ingenious experiments to show the characteristics of this cell, but the most pertinent for my argument here is the one in which he killed a founder cell with a hot needle. Immediately a number of outlying cells in the vicinity became rounded potential founder cells, one of which ultimately took over and apparently caused its rivals to disappear, that is, the other newly formed rounded cells become elongate again (essentially the same thing happens if a queen is removed from the nest of a primitive wasp—one of the other females takes over). This is clear evidence that the founder cell sends off an inhibitor of some sort which prevents other cells from turning into founder cells, and if the founder cell is killed the inhibition is released. Again the cells can be genetically identical (if they come from a clone started by a single cell, something easily done in the laboratory), which means that their dominance hierarchy is the result of somatic, physiological differences.

If we now turn to the stalked fruiting body, it is obvious that in the advanced dictyostelids the cells which form the stalk die and therefore are the equivalent of the sterile workers in social insects. They do not themselves produce offspring, but they contribute to the successful dissemination of their genes by supporting the fertile spores. How might this condition have arisen during the course of evolution?

An important sociobiological point is that slime molds can be either clones or a group of individual cells with different genotypes. This is also true for social insects. In the social

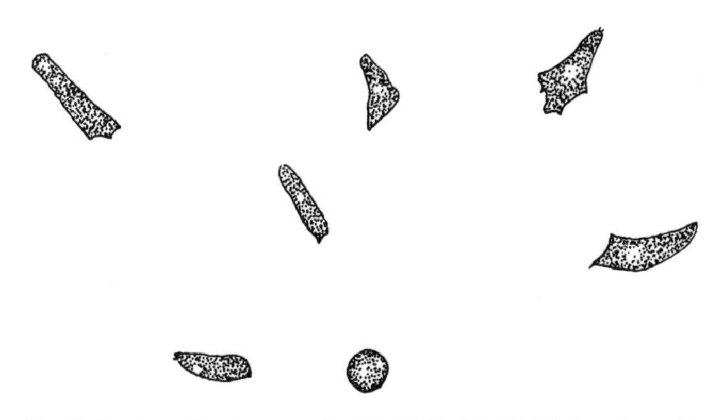

Fig. 7. Tracing of a photograph of B. M. Shaffer (1961) showing cells aggregating towards a founder cell in *Polysphondylium violaceum*.

hymenoptera, which have a sexual system of reproduction, there is a mixture of genotypes in the colony. In polymorphic aphids, which reproduce clonally in one phase of their life history, the genomes of each individual is identical (for a review, see Stern and Foster 1996). In the cellular slime molds, when there is more than one genetic variant in the soil, which has been unequivocally demonstrated by Francis and Eisenberg (1993) from soil samples in nature, then an aggregate can have a mixture of genotypes. However, it is a reasonable assumption that often some of the aggregates in nature will be clones, derived from single spores—clones similar to those aphid colonies that include a soldier caste. In the case of the clonal colonies of both slime molds and insects, the cost of producing a sterile caste must simply be more than compensated for by the benefit they confer to reproductive success. In those colonies where there is a mixture of genotypes, each amoeba or insect is out for itself, for its own genetic advantage, and the organisms' social coop-

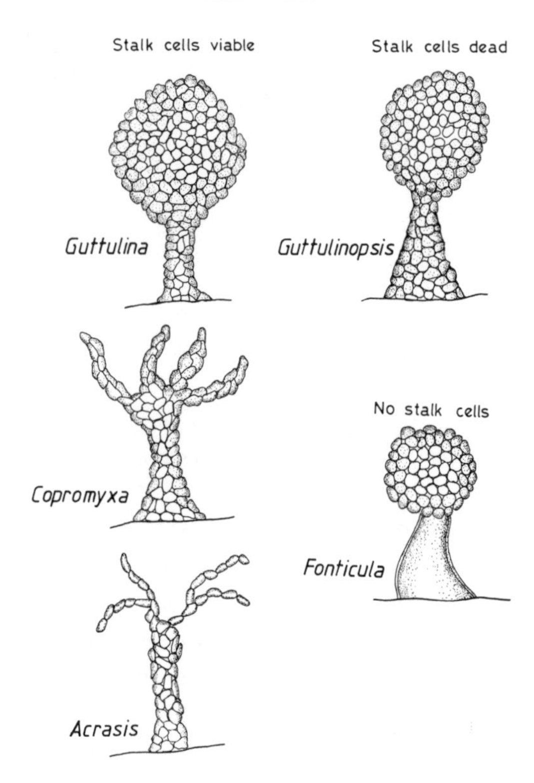

Fig. 8. Different genera of acrasids arranged so that one can compare those whose stalk cells are alive and capable of starting a next generation, and those with dead stalk cells. (From Gadagkar and Bonner 1994)

eration depends on the degree of their relatedness, as embodied in the Hamiltonian principle of kin selection.

An insight into the answer of the question of how sterility arose in the first place comes from a look at different species of acrasids (fig. 8). In acrasids the mound of aggregated amoebae form a small, irregular fruiting body that rises up into the air. They may have a simple terminal knob (*Guttu-*

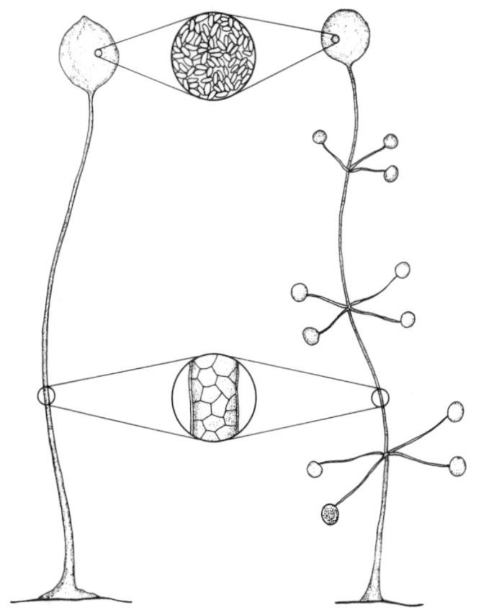

Fig. 9. A comparison of *Dictyostelium* and *Polysphondylium*. Inserts show the details of the cell structure of spores (*top*) and stalk cells (*bottom*). (From Gadagkar and Bonner 1994)

lina and *Guttulinopsis*) or they may be branched (*Copromyxa* and *Acrasis*). The interesting point is that in *Guttulina, Copromyxa,* and *Acrasis* any cell, even one from the stalk, can produce a live amoeba that can start the next generation. However, in *Guttulinopsis* the stalk cells have lost this ability; only the terminal "spores" are viable. This so exactly parallels what is found among primitive wasps that one can directly apply the sterility hypothesis of M. J. West-Eberhard mentioned earlier. Some of the stalk cells, by chance, lose their ability to propagate, but because they are not needed for successful dispersal, it makes no difference. Initially it is a division of labor by chance loss of the genes which ensures the ability of all the cells to reproduce; but then presumably there could be a selection to improve their role in supporting the spores and enhancing dispersal, ultimately producing the beautiful stalks found in the dictyostelids (fig. 9).

Indeed, in those latter species there is evidence that presumptive spore cells send off between-cell signals that inhibit the presumptive stalk cell from becoming spores. This again has its counterpart in social insects, where the reproductives will inhibit the workers from becoming fertile by means of a between-organism inhibitor. Work on the problem of how proportions are controlled in *Dictyostelium* is being actively pursued in many laboratories. It clearly involves a complex set of inhibitors and activators which are slowly revealing their secrets. To me the most interesting substance so far is a small molecule called DIF discovered by R. Kay and his co-workers (for a review see Williams 1991). It has the ability to induce cells to become sterile stalk cells, and it is produced in the presumptive spore region of the cell mass. It is a perfect parallel for the kind of manipulation found in the queens in social wasps. What I find most surprising is that the insect society castes seem to be controlled by relatively few pheromones, yet slime mold proportions involve a complex network of between-cell signals. As our knowledge of the control of division of labor increases for both the social insects and the social amoebae, perhaps we will be able to give an answer to this seeming paradox.

There is one further interesting point if one compares *Dictyostelium* with *Polysphondylium*. In the former there seems to be an extended preliminary period where all the between-cell signaling is operating—a kind of planning period where the signal network is in operation and organizing the final differentiation. Thus far there is no evidence for the equivalent preparatory period in the beautifully branched *Polysphondylium* (fig. 9). One has the impression that its proportions, which are strictly adhered to, have been cast in cement and there is little in the way of a preparatory period for their arrangement. This even extends to the founder cells, where again the development of *Polysphondylium* seems far more clearly defined than that of *Dictyostelium*. One possibility is that there is a more complete genetic control over each developmental step in *Polysphondylium*, something that might have occurred by gene accumulation.

Another Mechanism That Controls the Division of Labor in Insect Societies

Some recent interesting evidence indicates that there can be an additional factor, governed by natural selection, that contributes to the division of labor in insect societies. This stems from the work of D. M. Gordon (1995), who showed that in harvester ants each worker is endowed with the capability of performing numerous tasks, but what task a particular ant performs very much depends on what its worker neighbors are doing, which in turn is determined by the number of contacts it has with its neighbors. As a result the colony as a whole can divide the labor effectively simply by the interactions of the individuals that can perform the numerous tasks. Which task they perform at any one moment depends not only on environmental cues, but on cues from other workers. This has been modeled, and recently Pacala, Gordon, and Godfray (1995) have made a model that takes account of both kinds of cues. It has been shown by Gordon (and predicted by the model) that as a colony of harvester ants grows over the years and the number of workers increases, the frequency of individuals interacting will decrease due to the avoidance of one another when the population is dense. In this manner the following whole-colony properties emerge: the division of labor becomes appropriately apportioned to meet the environmental change, and the efficiency of doing this increases with colony size. All of these whole-colony responses are explained by the capabilities, or task repertoires, of the individual worker ants, and their sensitivity to the environment as well as to the activity of their sister workers. The optimality achieved in this democratic fashion is obviously favored by natural selection. Group behavior therefore can be understood in an elegant fashion in terms of the behavior of the similar individuals in the group.

Insect and Human Societies

I would now like to consider insect and human societies—not from the point of view of how they first achieved their division of labor, but to show that they embody the same principles that we have seen in the examples so far, namely, that there is a relation between size and division of labor, and that in all cases they can be accounted for by the selection for efficiency.

Insect societies are quite remarkable for their size and their diversity. Some ant colonies will have one queen and millions of sterile workers attending to all the activities within the nest. These consist primarily of feeding and caring for the young, foraging for food, and protecting the colonies. Some species of termites have equally large colonies, but they differ from ants, bees, and wasps in a few important respects, although these differences are not germane to my argument here. At the other end of the spectrum, the primitive wasps that were discussed previously may have very small colonies consisting sometimes of less than a dozen individuals.

In the large colonies of ants and termites there are morphological differences among the workers. They may not only differ in size, but in shape as well, and those differences are associated with specific tasks within the colony. The smallest workers may be primarily occupied with the care of the young, the middle size workers are busy foraging for food, while the largest workers, the soldiers, will guard and protect the nest. To see if one could measure the extent of this morphological division of labor in relation to the size of the colony, I took a table from B. Hölldobler and E. O. Wilson's (1990) great tome on ants, which gave the colony size for a large number of species. I broke this up into groups based on colony size, and Wilson was kind enough to indicate how many morphological castes there were for each of these species. He pointed out to me that for many it was easy, for the castes are discrete in their shape; but for others there is a continuum, and therefore assigning a number for different cases is difficult and subjective. I was fortunate to be able to rely on his great experience in observing ants and could plot colony size versus the estimated number of castes. I could show that, as expected, there is a clear and significant increase in the size of the colony, despite a considerable scatter of points (fig. 10).

In this case the explanation of why there should be such a relation is undoubtedly again natural selection. Insect colonies, with their one fertile and mated queen, pass all their genetic information from one generation to the next through

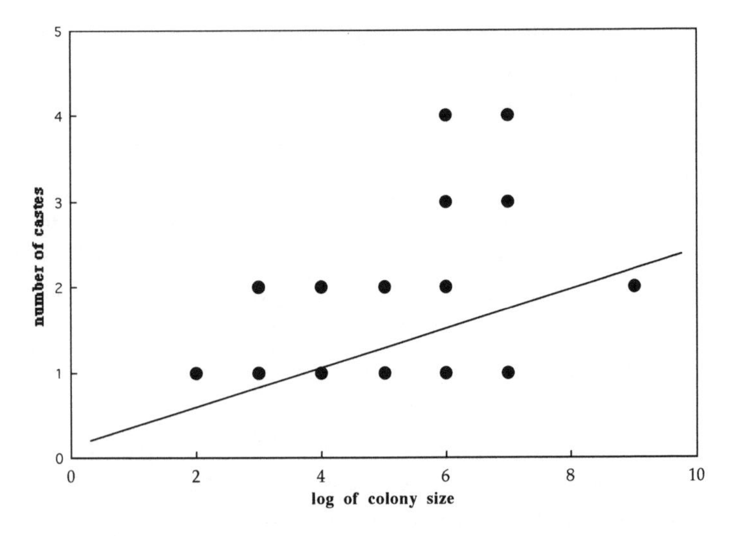

Fig. 10. The estimated number of castes of 140 different species of ants plotted against the logarithm of their colony size. The line is a simple regression: there is a significant correlation between the number of ants and the number of different castes in the colony. (From Bonner 1993)

her eggs. She produces all the sterile workers and the reproductives that start the next generation. If a queen carries genes which produce a large colony, clearly those colonies that divide the labor efficiently will survive. As the renowned entomologist W. M. Wheeler (1911) pointed out many years ago, a specialized worker in an insect society is equivalent to a specialized somatic cell in a multicellular organism. Only the reproductive cells of our body carry our genes on to the next generation, and all our other cells die at the end of our life cycle. The cells that divide our labor are in all senses the equivalent of sterile workers of different castes in an ant colony. For both ant societies and multicellular organisms, it is efficiency that leads to reproductive success, and that efficiency means an appropriate division of labor. The important point is that a complex insect society can easily be understood in the same way one explains the division of labor of cells within a multicellular organism: it has arisen by the natural selection of genetic variants, that is, it has a totally Darwinian explanation.

The perception that there is a division of labor in human societies goes back at least to Adam Smith, the eighteenth-century economist, and has been pursued by many others since then. Herbert Spencer, the nineteenth-century sociologist-philosopher, saw this, in his enthusiasm for generalizing, as an important principle of biology and sociology. His work implies that the division of labor is related to size, but there was no effort until recently to measure the increase in the division of labor with an increase in size of the society. It was simply pointed out that in very small communities individuals were expected to carry out more than one craft; but as a village increased in population, there could be a clear separation so that a cobbler made shoes, a tailor made clothes, a butcher sold meat, a greengrocer sold fruit and vegetables, and so forth. These special trades or crafts could only be separated into the activities of single individuals when the community became big enough, for example, to give a cobbler a full-time job. In a very small village he would have to do other things besides shoemaking to earn his bread.

These generalizations were put on a quantitative footing by R. L Carneiro (1967), an anthropologist who, building on some work of others, examined a number of societies of different sizes from different parts of the world. For each he measured the number of what he called "organizational traits," which was an attempt to identify specific crafts and occupations in a society, including those that contributed to its social structure. The smallest groups were formed among the now extinct Tasmanians, which consisted of around fifteen individuals and were differentiated into two organizational traits. The two largest groups in this study consisted of approximately fifteen hundred individuals, and possessed a total of thirty-two and fifty-two organizational traits. In all, he measured forty-six single-community societies, and when he plotted these on a log-log graph he was able to show a linear relation between the size of a community and its division of labor (fig. 11).

To give another example, M. Gadgil and N. V. Joshi have been working with K. S. Singh (1992) and his colleagues on the People of India project carried out by the Anthropological Survey of India, which is making a wonderfully rich collection of measurements of communities from all over India. Joshi very kindly retrieved for me from this great pool of data the population size of different Indian states and the number of occupations found in each. Again in a log-log plot one sees a clear relation: the larger the population, the greater the number of occupations, or division of labor (fig. 12). The smallest state has a population of 40,000 with a total of 13 occupations, while the largest has a population of slightly over 110,000,000 and 142 occupations.

Before asking the question of why this is so, I would like to consider another kind of human activity where one finds the same principle. This is something that exists in human organizations such as businesses, universities, government bureaucracies, and other similar institutions. There too, as they increase in size they increasingly divide the labor. Consider, for instance, a country store in a small village. It sells everything from food to clothing to tools to stationary.

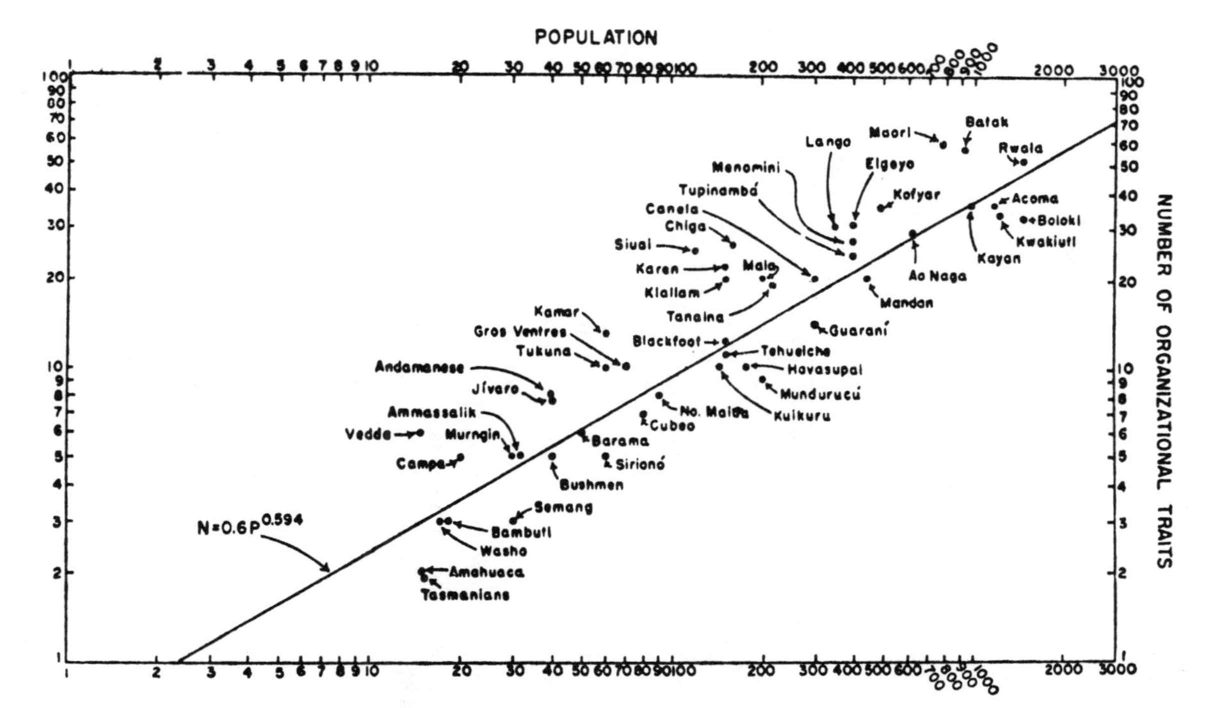

Fig. 11. The number of organizational traits plotted (log-log) against population size for forty-six single-community societies. (From Carneiro 1967)

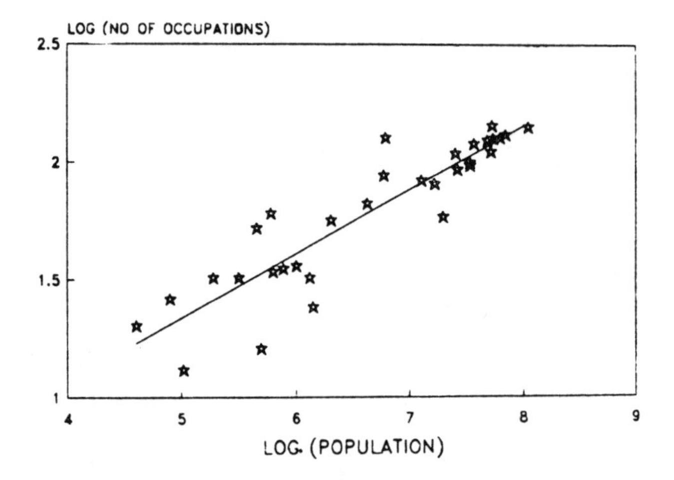

Fig. 12. The number of occupations found in each of the states in India plotted (log-log) against the population size of each state. (Graph by N. V. Joshi, in Bonner 1993)

They will not have a large stock of any of these items, but the variety will be sufficient to meet the major needs of the villagers. The larger village has a baker, a hairdresser, a butcher, a cobbler, and so on. It will even have a division of labor among the professions, so that there will be a doctor, a lawyer, and a banker to take care of one's more sophisticated needs.

It is easy to see that one finds the very same trends if one goes from small businesses to larger ones, or small universities to larger ones, or even to departments within a university or a business. With an increase in size these institutions generate a need for an internal structure that not only divides the labor, but does it so that there is good communication between the more numerous members and that the job of each does not overlap, but dovetails with all the others. This is obviously necessary for a business to be competitive and effective. If that is not achieved, the company will fail, and therefore at each size level there is an increasing but optimal division of labor, so the business, no matter what its size, can succeed.

In recent years some businesses have also undergone what might be called the "department store" phenomenon. Instead of remaining discrete and manufacturing one product, there has been a strong tendency for a company to diversify so that, like department stores, they are concerned with many commodities. This has even led to a merger of large companies into megacorporations. However, the principle that the division of labor increases with size holds equally well for all those amalgamations; their economic success depends upon the smooth inner working of all the parts, which must be effectively coordinated. The only place where efficiency is in danger is in the great government bureaucracies where the dividing line between success and failure is not so sharp.

If we now ask why human societies seem to follow the same size–division-of-labor principle of cells, the answer is somewhat different from the answer for cells. There are no obvious physical constraints, as we saw for a cell or for a multicellular organism; but even more important, there is a totally new element, namely behavior. In general, individual organisms can pass on information only through their genes, but in the evolution of animals there arose an entirely new method of passing information, and that is by behavioral signals from one individual to another. In order to contrast this behavioral transmission of information with genetic transmission, R. Dawkins (1976) has called the former "memes." A meme is a bit of information that can be passed from one individual to another by behavioral means.

The difference between memes and genes is very large, and the consequences of those differences are important. Genes can be passed on only through egg and sperm, from one generation to the next, and therefore many generations over a long period of time are required for a particular gene to spread through a population. Memes, on the other hand, can pass from one individual to another in an instant of time, and as a result they can spread rapidly through a population.

There is, of course, much more to behavior than the passing of information between individuals; we can also solve problems. Other animals also have the power to solve problems, but of all animals we are undoubtedly the most skilled in doing so. It is this problem-solving ability that is the key to how the labor is divided in societies.

First let us look again briefly at ant societies. I made the point that although their division of labor is somatically determined by environmental cues and chemical signals, the *capacity* for division of labor is primarily genetically based and therefore subject to natural selection. This is certainly true, but insects have behavior as well, and perhaps even a modest ability to solve problems. Those who work with social insects can give you many examples, but here I will just give one that is relevant. If, in an ant colony which has different castes performing different labors, one removes one of the castes, then some of the workers of other castes will take over and perform tasks they had never performed previously. In other words, they show a behavioral flexibility, one of the elements needed for problem solving. By being flexible they see to it that all the functions, all the labors required for the welfare of the whole colony are carried out, despite the loss of one of the castes. From this one might conclude that in social insects there can be a mixture of behavior in a background of a strong genetic basis for the division of labor.

For human beings the genetic component takes the back seat. No doubt the structure and abilities of our brain have primarily a genetic basis, but what we do with those skills opens up a new world.

With this background let us now examine the size–division-of-labor structure of human societies. If we look at the relation between occupations and the size of communities, it is clear that for the economic survival of the individuals within the community they must follow a course that is dictated by the size of the community. If they fail to do so, those individuals may starve and immediately the whole community will be jeopardized. But human beings are supreme problem solvers, and as I pointed out earlier, if there are not enough people in the community to keep the shoemaker occupied full time, he will take on other work as well and in one person combine more than one occupation.

If we compare this to natural selection, it is clear there is also a selection in human societies. The principal and immediate selection pressure, however, does not come from reproductive success, but rather from economic survival or economic stability. In the world of nonhuman animals and plants, survival is also crucial. If an animal starves it will not reproduce, so in this sense there is no difference between the selection of human beings and other organisms. But in another sense there is a great difference, for human beings rely heavily on behavioral or cultural means of information transmission. So the success of any one human being depends upon his or her awareness of the immediate situation, the immediate environment, and the ability to manage economically and make a way in the world. In other words, the process of survival in human societies has become somewhat removed from reproductive success. The human mind not only can solve problems, but it can even make arbitrary decisions which may or may not foster reproductive success. In fact, often such decisions reduce reproduction, such as electing celibacy for religious reasons. Our brains have allowed us a great variety of different kinds of behavior that affect our social structures. They may primarily involve problem solving, but they also might be present because of traditions, or superstitions, or through purely thoughtless actions. The significant point is that where there is human behavior, reproductive success has taken a secondary role because it is overshadowed by all the remarkable mental activities that are within our command. We cannot escape natural selection, for we have genes and we do reproduce, and there is ample evidence that natural selection can and does affect the genetic constitution of our populations. It is simply that concern for genetic continuity has all but disappeared in the mind of *Homo sapiens* and problem solving and other mental activities have become all-consuming occupations.

An important component of our search for stability, and even success, is to maintain, through the powers of our brain, a great effort for maximum efficiency. And one important component of efficiency is a division of labor within the community in which one lives. Moreover, as the community changes in size, the requirements for efficiency change, and shifts in the extent of the division of labor become important. In other words, I am arguing that for human societies and human institutions, the most efficient division of labor needs to be optimized by our problem-solving minds. It may even be that we achieve a reasonable balance of division of labor by trial and error, but that is simply one of the ways we solve problems, as every experimental scientist knows. By contrast, nonhuman animals and primitive multicellular organisms display a different kind of trial and error. It involves the production by random mutation of variants that affect the division of labor, and those that are most efficient will be passed on to the succeeding generations and ultimately fixed in the population. Again we have distanced ourselves from our genetic selection by our extraordinary ability to use our mental powers and to solve problems by behavioral means.

Conclusion

My thesis has been that even though it is impossible, for purely mechanical reasons, for any large organism to avoid having a division of labor, it is not something that appears spontaneously, but both its origin and the form in which it appears are driven by selection. As for the origin, we see clear examples of its early appearance in *Volvox*, in cellular slime molds, and in social wasps, and in each case the argument that natural selection is the guiding principle is both reasonable and compelling. That selection also plays a part in the continuation and perfection of the division of labor has not been equally stressed, but it is obvious that many of the engineering problems imposed by size increase during the course of evolution have been solved by a variety of different mechanical methods. There is no law which says how the labor is divided; only that such a division is a physical necessity. The division itself is made by the trial and error of natural selection, and the devices that survive are the ones that lead to reproductive success.

It is particularly interesting to note that initially all division of labor seems to involve somatic variation among the parts, be they cells or individual organisms. When the division of labor becomes more organized, a genetic component creeps in. This genetic component is primarily in the form of a control over the extent and the nature of the somatic variation. A prime example is the appearance of sterile workers in social insects or sterile stalk cells in slime molds. Somehow in those organisms the ability to manipulate environmental cues so that some cells or individuals will be sterile is part of the genetic constitution of all the members of the social group. Every individual in the colony has the set of genes that controls the extent and nature of the somatic variability, including the way it responds to environmental cues.

Finally, we see that the division of labor that is so pervasive in human society does not have the physical constraints found in cells and in individual multicellular organisms, and is even farther removed from genetic control. Yet clearly there has been a selection for efficiency, and that efficiency cannot be achieved without a division of labor.

CHAPTER 5

Sixty Years of Biology

My life as a biologist has spanned sixty years, and the changes during those years have been staggering. I was there to watch its many metamorphoses unfold right in front of my eyes. Among the milestones there is one overall trend that is striking. When I began, most of biology was about big things. For instance, if one follows the course of how we have thought about Darwinian natural selection, we see that in the 1930s evolutionary biologists were involved with whole populations; later natural selection was applied primarily to individual organisms, and even more recently to their genes. Although I am a firm believer in looking at the living world in a grand, overall, holistic manner, it is clear that in our great progress in most of the fields of biology during these three score years there has been a simultaneous progressive spreading into reductionism. Many of the answers we have sought and found have involved explanations at a lower level of analysis; some great successes have come from atomizing biology. Even though this is where much of the excitement lies, we can only appreciate and fully understand the great lessons of this reductionist revolution and its applications to all of biology, including applied fields such as medicine, if we examine how the reduced parts fit together to make the beautiful whole.

CHAPTER 5

I first became deeply interested in biology in 1932 when I was twelve years old. My parents were temporary American-expatriots and we were living in London in a house at Onslow Gardens, not far from the Museum of Natural History. During the school holidays I spent many hours poring over the exhibits. The birds and everything else in the museum seemed enormously exciting. It was my first flush of exhilaration that goes with discovery. I am happy to report that in many ways I have changed, and stuffed birds and bones no longer have the same grip on me that they did in those early years. Fortunately, something new and important cropped up more or less continuously and kept me in a permanent state of excitement, just watching the progress of the science of life unfold with many starts and great spurts. With a broad brush let me outline the stupendous changes that have occurred during my own biological lifetime.

More than any area of biology, genetics has shown the most spectacular changes of all. Gregor Mendel published his famous paper in 1866, but it was not "discovered" until 1900. It was not appreciated until the early part of this century that the chromosomes within the nucleus contained the genes, the factors of Mendel. Then T. H. Morgan and his equally gifted collaborators at Columbia University were able to understand in detail how the genes were organized on the chromosomes, and how they could be rearranged and exchanged during the process of gamete formation and fertilization. Morgan used *Drosophila*, the fruit fly, for his experiments, for they have a short generation time, and he raised them in the laboratory in milk bottles. My first encounter with this work was in 1933 while I was at school in Switzerland when our class took a trip to Geneva to visit the university. We went to an open house in Professor Guyénot's laboratory, where the professor was following some ramifications of Morgan's experiments on fruit flies. He was a smiling, courtly man, wearing a beret, who obviously delighted in showing these children his flies and telling them what happened when one bred white-eyed flies with the normal red-eyed ones. In the summer of 1938, after my freshman year in college, I took a course at the Marine Biological Laboratory in Woods Hole, Massachusetts. I knew that Professor Morgan, now famous and the grand old man of genetics, was there, and that a friend of my parents was his nephew. With considerable hesitation I asked for a letter of introduction to his uncle, and with even greater trepidation I called on Professor Morgan one evening. He and Mrs. Morgan could not have been kinder to what must have seemed to them a rather pathetic, aspiring teenage biologist. I was thrilled at the time and I know the visit had many good effects on me, but it did not steer me towards genetics.

From the discoveries of the Morgan school came an enormous flowering of genetics, and by the time I was teaching as a beginning assistant professor the crucial question became, What was the chemical nature of the genes themselves? We all know now that it is DNA, but in the early 1950s it was a subject of hot debate, and many people, wrongly of course, favored proteins over DNA.

The greatest single biological revolution during my life was certainly the well-known discovery of Watson and Crick of the structure of DNA. It now immediately became clear how DNA could make template copies of itself in its double helix. This earth-shaking discovery was quickly followed by many others: how the DNA coded for specific proteins, and all the steps that carried out the process. Ultimately there was a detailed understanding of exactly how genes gave their orders. These discoveries involved many people (and many Nobel prizes)—it was a triumphant procession of staggering significance. All of molecular biology today is the direct result of these revelations: they led us to genetic engineering, to being able to identify criminals from traces of their blood and to new ways of figuring out the ancestry or relatedness (family trees) of all sorts of different animals and plants simply by the magnitude of the differences in the details of the structure of their genes. They also have led, and will continue to lead, to enormously important medical advances, not only for genetic diseases but for combating many other diseases as well. And all this fantastic progress has been in the last twenty-five years! In the early days there were just a few geneticists like Morgan and Guyénot spotted here and there, and today they have been supplanted by vast armies of molecular geneticists all over the world. Even in the lowly slime molds, which have been the passion of my biological life, the majority of the exciting work today is on their molecular biology, which brings me directly to the subject of developmental biology, or what used to be called embryology.

The foundations of modern developmental biology began in the nineteenth century and the beginning of the twentieth. First there were the detailed descriptions of the embryology of many animals that led, both in Germany and in America, to the beginning of experimental embryology, which were the first attempts to understand the "mechanics" of development. A gifted and colorful German named Hans Dreisch showed that if one cut a sea urchin embryo in two, down the middle, both halves turned into minute but perfectly proportioned larvae. (He was so astounded by this result that he decided it could not be explained by mechanical causes—there must be a vital spirit!) In America, E. B. Wilson and E. G. Conklin showed that embryos of other invertebrates were very different, and each half-embryo produced half a larva. Conklin was a very emeritus professor at Princeton when I first came there to teach, and he loved to reminisce how he and Wilson worked on different, unrelated animals, and how one Sunday morning at Woods Hole they compared notes to find to their utter amazement that the developments of their beasts were identical. By the time I was a university student it was appreciated that both kinds of development exist, and, furthermore, generally all organisms have a mixture of the two, some emphasizing one more than the other. Perhaps the most important experiment of all was that of H. Spemann, who showed that a particular portion of an amphibian embryo, when transplanted, induced a second embryo. Even though these experiments were done before I was born, their impact is felt to this day because they demonstrated that there are chemical signals that control the pattern, and modern developmental biology has centered around understanding those signals. This is the kind of developmental biology that has kept me busy over the years—looking for signals in slime molds. It has been actively pursued in animal embryos, and there has been great success in analyzing the chemical signal systems in plant development. Recently it has been possible to analyze the genes that are responsible for the production of the chemical signals, and the ones that are responsible for the responses of the cells to those signals. This has produced a great new wave of excitement. Fruit flies, nematode worms, and (of course) slime molds are being vigorously attacked and are giving up their secrets. Here is the meeting ground of genetics, molecular biology, and development—a great new horizon has opened up.

One of the central subjects of the nineteenth century was cell biology, culminating in E. B. Wilson's great book *The Cell in Development and Heredity*, first published in 1896. The summer I met Morgan I remember seeing Wilson walking about on the street in Woods Hole, but I was too ignorant then to realize who he was—a missed opportunity I have always regretted. The study of the cell has also made enormous advances during my lifetime. The first wave was the result of the rise in biochemistry: it became possible to study many of the chemical reactions that were occurring inside cells. This led to a deep understanding of metabolism, the chemical machinery that supplies the cell with energy for all its activities. Further, it led to an understanding of the structures of the cell: the membranes, the chromosomes, the spindle fibers, and the numerous fine structures that were revealed by the invention of the magic electron microscope, which was effectively put to use in the 1950s. The second major wave has again come from molecular biology, so that not only can one trace the different chemicals that pass from one part of the cell to another, but one can analyze the molecules involved through their genes which govern their structure and their activities. My only reservation about all these recent wonders in molecular and cell biology is that the more we know, the more elaborate and complex living organisms seem to be. Fortunately, now and then a new insight emerges that generalizes and simplifies the relation between the innumerable parts of the cell or a developing embryo.

Neurobiology and the behavior of animals are other areas of biology that have leapt forward during my career. It was during this period that A. L. Hodgkin and A. F. Huxley elucidated how nerves transmit impulses, an enormous advance that provided a bridge on how to relate the activity of a single nerve with that of a whole nervous system, including a brain—a subject hotly pursued in many laboratories today. Animal behavior is a venerable subject, but it suddenly took on a radical new life with the ideas and the wonderfully ingenious experiments of K. Lorenz, N. Tinbergen, and K. von Frisch. They showed that animals have specific responses to specific stimuli (just as in development), and they were able to make the notion of "instinct"—which had been banned from our vocabulary when I was a student—respectable and acceptable. There are innate responses and learned responses. This not only led to advances in our understanding of behavior, but made it possible to ask genetic questions about behavior, another subject that always flirted at the borderline of a tabu. Furthermore, animals, even bees, could show remarkably complex behavior, as von Frisch showed in his beautiful experiments.

Lorenz came to Princeton to give a lecture some years ago and he was a wonderful showman. His lecture was without doubt a marvel, full of bird calls and bird postures, along with a wonderful grasp of the mood of his audience. The next day a colleague and I took him to our "perception center" in the psychology department, which had a series of rooms, each of which illustrated an optical illusion. For instance, with one eye one could peep through a hole, and if two people of equal size were in two corners of the room, one seemed a giant, and the other a dwarf. When we arrived there, it turned out that Niels Bohr, the famous physicist, was going to make the tour at the same time. Introductions were made all around, and Bohr and Lorenz were like two excited children, each looking through the peepholes or standing in corners, having the most wonderful time. After we finished, Bohr asked if all of us could have some coffee together and discuss what we had seen. This suggestion was enthusiastically approved, and as we sat around in a circle, Bohr began to talk, that is to say, he mumbled terribly in a thick Danish accent—he was exceedingly hard to follow. As we struggled along, Lorenz, a brilliant talker, kept trying to say something too, but Bohr ignored him totally and went on serenely with his "discussion." Poor Lorenz was beside himself with frustration. After a half hour the monologue was over, and none of us had gotten a word in edgewise. It was fascinating to watch the two egos clash, each with his own technique of dominating a conversation. In this case it was a knockout by Bohr. Later I tried to put together in my own mind what Bohr had said, and it dawned on me that his message was: things are not always the way they seem.

During the last sixty years there has been another set of important advances in ecology and evolution and related fields. My first encounter with ecology was the result of a traveling fellowship I received after graduation from college in 1941. I spent part of the summer on Barro Colorado Island in Panama—a natural island preserve that was created by the flooding of Gatun Lake when the canal was built. It was my first introduction to a tropical rain forest—and I was overwhelmed. I spent most of the day out in the wild, utterly staggered by the profusion of animals and plants. The only drawback was that I was alone much of the time, I had to do my learning without a teacher. I spent the evenings reading Proust, which was perhaps a unique way to learn tropical ecology. For a few days there was a visitor—a distinguished ecologist of the old school who taught me some things—but we spent more time arguing. He believed that the important thing about nature was its complexity, and we could learn nothing from experiments; that was interfering with nature. I remember the high point came when he told me that all of genetics was bosh because it was done in milk bottles, which reminded him that in his youth he had earned money as a milkman! I was quite polite about it all, despite a seething, youthful outrage.

Largely because of this episode I might have remained quite ignorant all my life of the beauties of ecology had it not been for Robert MacArthur, who came to Princeton in the 1960s. His entire approach had its origins under the encouragement of his mentor, G. E. Hutchinson. By the time he joined our department he was already a major figure in the revolution in ecology in using mathematical models to get insights into deep ecological mechanisms. It was possible for him to do this because he was a master at field work, and with good judgment he fused his knowledge of natural history with his skill in mathematics. He was bitterly attacked by the old guard for trying to simplify the very complexity in which they gloried. (His answer was, Where would physics be without frictionless pulleys?) We became good friends, which made his tragic death from cancer when he was only forty-one years old a devastating blow. But his was a revolution too, and much of modern ecology can be traced back to his approach. In recent years ecology has become a mature science in which one can be interested in natural history yet ask profound questions about how the environment is put together, something that is now vital in our new concern with conservation. Happily we are way past the days of "thick description."

Evolutionary biology arrived as a volcanic eruption through the work of Charles Darwin in the last century. Its history before and since Darwin has been fascinating to a large extent because it is concerned with many of the same questions as religion. They are questions of how and why we got here and why we are the way we are—questions that will forever produce a tug of war between science and religion. Even the science part has not been without controversy and big changes, many of them in my lifetime. The number of biologists who were satisfied that natural selection could account for evolution was pitifully small from Darwin's time up to the 1930s. Most often the view was expressed that selection could account for the degeneration, or the removal of undesirable traits, but not for the appearance of new characters. For that there had to be some sort of "vital spirit"—divine or secular.

The tide turned with the rise of population genetics, in which R. A. Fisher, J.B.S. Haldane, and S. Wright used mathematics to show how selection and other factors which lead to change could alter the frequency of individual genes in a population. This new approach was somewhat grandly called "the new synthesis," and T. Dobzhansky, A. H. Sturtevant, E. Mayr, and others did much to expand it to questions of how new species arise and to other global evolutionary problems. Most of these pioneers were active for many years of my life, and indeed Ernst Mayr is still going strong. In my senior year at Harvard I took a course on this very subject taught by Sturtevant, using the book of his arch rival Dobzhansky as the text.

Yet of all these biologists, Haldane was perhaps the most original, and certainly the most wide ranging. One of his conspicuous qualities was that he loved to shock, especially when dealing with authority or establishment, and he put on a good show. Once, in the late 1950s, he asked me out to dinner with his wife Helen Spurway. A friend took me to one side and said, "You will go to a sleazy-looking restaurant in Soho, but the food will be good. Do not be put off by the fact that they will discuss some aspect of sex in very loud voices and people at the neighboring tables will stare at you." That is exactly what happened, but my hosts had thoughtful things to say about everything, including sex. I never met R. A. Fisher, but have heard him lecture at Princeton. I remember he was very difficult to follow and would deliver whole paragraphs staring at a piece of chalk which he held about two inches from his face.

In recent years the interest in evolution continues to rise. As I said earlier, in part this is due to molecular techniques for plotting the ancestry of animals and plants. A more important reason is that ecologists see that their concepts must be understood within a framework of evolution. Also, there is an increasing interest today in the old idea that it is not just the adult that evolves, but the whole life cycle, including the organism's development. C. H. Waddington was an embryologist who became interested in these problems and did much to further the idea. I spent a sabbatical in his laboratory in 1958 and benefited from it. He was an intense man of tremendous drive and ability. I have many pleasant recollections from that period in Edinburgh, with my growing family. We liked it so much that I seriously considered trying to find a way to stay.

Without doubt the biggest revolution in the study of evolution in recent times—one of Nobel proportions—was the insight of W. D. Hamilton. Working with social insects, such as worker ants or bees, he realized that if individuals within a social group help one another, and if they are genetically related, those genes they share will be passed on to the next generation, even if some of the individuals are sterile. This led him and others to realize that the genes themselves are selected (as R. Dawkins describes so well in *The Selfish Gene*). This insight has led to a much clearer understanding of why so many animals are social, for one of the advantages of togetherness is genetic. The result has been, since the mid 1970s, a great surge in the study of sociobiology, which E. O. Wilson helped bring to the attention of the world in his fine book, *Sociobiology*.

In this brief sketch of how Darwinism has matured in the last sixty years lies a profound message. The early population geneticists were interested in populations and how the frequency of genes changed in those populations. In the new synthesis these ideas were used to explain how new species came into being. The whole question of how selection was acting on organisms came to a boil in the 1960s when V. Wynne-Edwards advanced the idea that natural selection could act on a group, and this stimulated a strong counterargument that the most important "unit of selection" was the individual organism. This was followed by Hamilton's idea that the genes themselves are the ultimate units of selection, and organisms are simply "vehicles," to use Dawkins's felicitous term, to carry the genes. Look at the matter backwards: the genes in our bodies are the ultimate survivors (although that says nothing about what will happen to a particular gene in the future). The organisms, the vehicles that carry those genes, come and go each generation.

My own particular obsession has been to emphasize the important point that these vehicles are not simply adults, but life cycles, and the genes carried in the life cycle play roles in governing during early development, in maturing, and even during senescence. The reason for life cycles is that sexuality—so essential for, and selected by, natural selection to control the changes, the variations in genes—requires a single-cell stage, the fertilized egg. This is the only way it is possible to produce a new organism that has the genes from both the father and the mother. If the genes are to survive unto the next generation, they must control the construction of an effective vehicle to ensure their perpetuation. Selecting life cycles and the genes that control them has led to the production of molds, of insects, of worms, of grasses, of giant sequoias, of elephants, and of all the millions of species that surround us in the world today.

To me the most fascinating thing is that in worrying about the way natural selection works, we have remained holists while at the same time becoming ever more reductionist. Darwin's original idea coupled to genetics started off an essentially atomic way (the genes being the "atoms") of understanding evolution. First, the genetic composition of populations was the important issue; then later the center of attention shifted from groups to individual organisms, or life cycles; and then finally to the selfish genes. In recent years this has produced a debate as to what are the "true" units of selection. Like so many arguments in science, everyone is right. The reason is that all three levels are equally important; one cannot exist without the others. It is quite conceivable that one species can survive or disappear with major changes in the environment—in fact, the extinction of species is happening all the time. Populations, or groups of one species, can quite easily thrive or disappear for the same reasons. The individuals, as Darwin understood so clearly, are obviously objects of selection. Furthermore, it has been pointed out by Leo Buss (1987) that there can be cell lineage selection within a multicellular organism. Finally, the competition among genes is clear and obvious and manifests itself in many ways. People don't like to be told everyone is right—they prefer to argue. The only sense in which genes can be said to be particularly important units of selection is that they are the ultimate ones—they are the "atoms" of evolution. But they cannot operate without all the other hierarchical levels of which they are a part; in this sense, evolution is a holistic enterprise. Everyone wins.

There has been another fortuitous outcome of the progressive reductionism of modern biology. The more we examine and discover the chemical nature of signal molecules, enzymes, and especially the DNA sequences of genes, the more we see that instead of the details of the molecular structure of organisms multiplying indefinitely, there are certain basic molecular communalities that are found in all animals and plants. The same chemicals serve as signals, as enzymes, and as other key proteins, including the genes that designate those proteins, in all organisms, from bacteria to the largest and most complex living beings. Reducing everything to biological microunits has not always been a source of confusion, but often one of enlightenment, for one can reach important and encompassing generalizations from the microstructure—in this case, biological reductionism seems to be slowly sneaking up to produce a new holism.

An equally dramatic change in biology and medical research has occurred in how science is done. In the fascinating biography *Darwin*, A. Desmond and J. Moore (1991) point out that in the early nineteenth century, pursuing science was only possible for those who had leisure time, Darwin himself being a perfect example. Later in the century, one began to see the first professional biologists such as Joseph Hooker, the director of Kew Gardens, and T. H. Huxley, a professor at the Royal School of Mines. In my lifetime the amateur scientist has almost disappeared—we are all professionals. There is also a striking new trend over my years on how we do science. When I began, the equipment we needed was modest, while now it is impossible to do many different kinds of laboratory biology without elaborate equipment such as electron microscopes, the new confocal microscope, complex spectrographs, high speed centrifuges, gas chromatographs, computers, and many more machines—each costing huge sums of money.

There is another big change in the way science is done, which comes from there being so many scientists. Many (but not all—I am not one of them!) experimentalists feel that it is not possible to progress fast enough unless one has a large group, and now publications have an increasing number of authors on their masthead. Despite the proliferation of new journals, it has become increasingly hard to find a place to publish. One reason for all this is that there are far more biologists today, and what with the big equipment, more money is needed than ever before to keep a large laboratory afloat. This has meant an enormous competition for funds, whose availability is not expanding at the same rate, with the distressing result that most biologists spend a large amount of their time writing grant proposals, trying to make them as earthshaking as possible to succeed in getting funded.

When I came to Princeton in 1947, all I asked for was a low- and a high-power microscope and the basic materials to culture my slime molds—there was no need for the "starter grants" that we routinely give beginning faculty today. A few years later the National Science Foundation was established, and I applied for what today would be considered a ridiculously small grant, and in the letter that told me I was successful they asked for an annual report in the form of a letter. After the first year I wrote that things had not worked out very well—I had tried this, that, and the other, and nothing had really worked. (Can you imagine writing such a letter today?) They wrote back saying, "Don't worry about it—that is the way research goes sometimes. Maybe next year you will have better luck." (Can you imagine the NSF writing a letter like that today?) So with all the wonders and marvels of our progress in laboratory biology during the last fifty years, there is a price we have had to pay. But for many kinds of experiments it could not be any other way. It may be fun to reminisce about the good old days, but it is far more rewarding to admire the truly extraordinary changes of the last sixty years.

BIBLIOGRAPHY

Ackert, J. E. 1916. On the effect of selection in *Paramecium*. *Genetics* 1: 387–405.

Barth, L. G. 1940. The process of regeneration in hydroids. *Biol. Rev.* 15: 4405–4220.

Basalla, G. 1988. *The Evolution of Technology*. Cambridge University Press, New York.

Bell, G. 1982. *The Masterpiece of Nature*. University of California Press, Berkeley.

Bell, G. 1985. The origin and early evolution of germ cells as illustrated in the volvocales. In H. Halvorson and A. Monroy, eds., *The Origin and Evolution of Sex*, pp. 221–256. Alan R. Liss, New York.

Bock, F. 1926. Experimentelle Untersuchungen an kolonienbildenden Volvocaceen. *Arch. Protistenk.* 56: 321–356.

Bonner, J. T. 1958. The relation of spore formation to recombination. *Amer. Nat.* 92: 193–200.

Bonner, J. T. 1965. *Size and Cycle*. Princeton University Press, Princeton, N.J.

Bonner, J. T. 1974. *On Development*. Harvard University Press, Cambridge, Mass.

Bonner, J. T. 1980. *The Evolution of Culture in Animals*. Princeton University Press, Princeton, N.J.

Bonner, J. T. 1988. *The Evolution of Complexity*. Princeton University Press, Princeton, N.J.

Bonner, J. T. 1993. Dividing the labour in cells and societies. *Current Sci.* 64: 459–466.

Brown, J. L. 1975. *The Evolution of Behavior*. W. W. Norton, New York.

Buss, L. W. 1987. *The Evolution of Individuality*. Princeton University Press, Princeton, N.J.

Byrne, R., and A. Whiten, eds. 1988. *Machiavellian Intelligence*. Oxford University Press, New York.

Carneiro, R. L. 1967. On the relationship between size of population and complexity of social organization. *Southwestern J. Anthropol.* 23: 234–243.

Child, C. M. 1941. *Patterns and Problems of Development.* University of Chicago Press, Chicago.

Connor, R. 1992. Dolphin alliances and coalitions. In A. H. Harcourt and F.B.M. de Waal, eds., *Coalitions and Alliances in Humans and Other Animals*, pp. 415–443. Oxford University Press, New York.

Cox, E. C. 1992. Modelling and experiment in developmental biology. *Current Opinion in Genet. and Evol.* 2: 647–650.

Cox, P. A., and J. A. Sethian. 1985. Gamete motion, search, and the evolution of anisogamy, oogamy, and chemotaxis. *Amer. Nat.* 125: 74–101.

Curio, E., V. Ernst, and W. Vieth. 1978. Cultural transmission of enemy recognition: One function of mobbing. *Science* 202: 899–901.

David, C. N., and R. D. Campbell. 1972. Cell cycle kinetics and development of *Hydra attenuata*, I. Epithelial cells. *J. Cell Sci.* 11: 557–568.

Dawkins, R. 1976. *The Selfish Gene.* Oxford University Press, New York.

Deneubourg, J. L., and S. Goss. 1989. Collective patterns and decision making. *Ethol. Ecol. and Evol.* 1:295–311.

de Waal, F. 1982. *Chimpanzee Politics.* Harper and Row, New York.

Desmond, A., and J. Moore. 1991. *Darwin.* Michael Joseph, London.

Desnitski, A. G. 1992. Cellular mechanisms of the evolution of ontogenesis in *Volvox. Archiv. Protistenkd.* 141: 171–178.

Edelstein-Keshet, L. 1988. *Mathematical Models in Biology.* Random House, New York.

Elsasser, W. 1966. *Atom and Organism.* Princeton University Press, Princeton, N.J.

Elsasser, W. 1975. *The Chief Abstractions of Biology.* American Elsevier, New York.

Francis, D., and R. Eisenberg. 1993. Genetic structure of a natural population of *Dictyostelium discoideum*, a cellular slime mold. *Mol. Ecol.* 2:385–391.

Gadagkar, R., and J. T. Bonner. 1994. Social insects and social amoebae. *J. Biosciences* 19: 219–245.

Gilliard, E. T. 1963. The evolution of bower birds. *Sci. Amer.* 209 (August): 38–46.

Gordon, D. M. 1995. The development of organization in an ant colony. *Amer. Sci.* 83: 50–57.

Gould, J. L., and P. Marler. 1987. Learning by instinct. *Sci. Amer.* 255 (January): 74–85.

Greenwald, I. 1990. Cell-cell interaction in the nematode *Caenorhabditis elegans*. *Curr. Opinion Cell Biol.* 2: 986–990.

Griffin, D. R. 1992. *Animal Minds*. University of Chicago Press, Chicago.

Hall, B. K. 1992. *Evolutionary Developmental Biology*. Chapman Hall, London.

Harrison, L. G. 1993. *Kinetic Theory of Living Pattern*. Cambridge University Press, New York.

Hickey, D. H. and M. R. Rose. 1988. The role of gene transfer in the evolution of eucaryotic sex. In R. E. Michod and B. R. Levin, eds., *The Evolution of Sex: An Examination of Current Ideas*, pp. 161–175. Sinauer Associates, Sunderland, Mass.

Hölldobler, B., and E. O. Wilson. 1990. *The Ants*. Belknap Press of Harvard University Press, Cambridge, Mass.

Hurst, L. D. 1992. Intragenomic conflict as an evolutionary force. *Proc. Roy. Soc. London B* 247: 135–140.

Jennings, H. S. 1920. *Life and Death, Heredity and Evolution of the Simplest Organisms*. Gorham Press, Boston.

Kazmierczak, J. 1981. The biology and evolutionary significance of Devonian volvocaceans and their Precambrian relatives. *Acta Palaeont. Polonica* 26: 299–337.

Kelland, L. J. 1977. Inversion in *Volvox* (Chlorophyceae). *J. Phycol.* 13: 373–378.

Kimura, M. 1994. *Population Genetics, Molecular Evolution, and the Neutral Theory*. University of Chicago Press, Chicago.

Kirk, D. L. 1988. The ontogeny and phylogeny of cellular differentiation in *Volvox*. *Trends in Genetics* 4: 32–36.

Kirk, D. L., M. M. Kirk, K. A. Stamer, and A. Larson. 1991. The genetic basis for the evolution of multicellularity and cellular differentiation in the volvocine green algae. In *The Unity of Developmental Biology*, pp. 568–581. Dioscorides Press, Portland, Oregon.

Leach, C. K., J. M. Ashworth, and D. R. Garrod, 1973. Cell sorting out during the differentiation of mixtures of metabolically distinct populations of *Dictyostelium discoideum. J. Embryol. Exp. Morphol.* 29: 647–661.

Maynard Smith, J., and E. Szathmáry. 1995. *The Major Transitions in Evolution*. W. H. Freeman, New York.

McDonald, S. A., and A. J. Durston. 1984. The cell cycle and sorting behaviour in *Dictyostelium discoideum. J. Cell Sci.* 66: 195–204.

McMahon, T. A. 1973. Size and shape in biology. *Science* 179: 1201–1204.

Medawar, P. B. 1952. *An Unsolved Problem in Biology.* London, H. K. Lewis.

Meinhardt, H. 1982. *Models of Biological Pattern Formation.* New York, Academic Press.

Michod, R. E., and B. R. Levin, eds. 1988. *The Evolution of Sex: An Examination of Current Ideas.* Sinauer Associates, Sunderland, Mass.

Murray, J. D. 1989. *Mathematical Biology.* Biomathematics Series, ed. S. A. Levin. Springer-Verlag, New York.

Nanjundiah, V., and S. Saran. 1992. The determination of spatial pattern in *Dictyostelium discoideum. J. Biosci.* 17: 353–394.

Needham, J. 1965. *Science and Civilization in China*, vol. 4, part 2, *Mechanical Engineering.* Cambridge University Press, New York.

Newman, S. A. 1992. Generic physical mechanisms of morphogenesis and pattern formation as determinants in the evolution of multicellular organization. *J. Biosci.* 17: 193–215.

Newman, S. A. 1994. Generic physical mechanisms of tissue morphogenesis: A common basis for development and evolution. *J. Evol. Biol.* 7: 467–488.

Olive, L. S. 1975. *The Mycetozoans.* Academic Press, New York.

Pacala, W., D. M. Gordon, and H.C.J. Godfray. 1995. Effects of social group size on information transfer and task allocation. *Evol. Ecol.* (in press)

Page, R., and G. E. Robinson. 1991. The genetics of division of labor in honey bee colonies. *Adv. Insect. Physiol.* 23: 117–167.

Petroski, H. 1992. The evolution of artifacts. *Amer. Scient.* 80: 416–420.

Pocock, M. A. 1933. *Volvox* in South Africa. *Ann. S. African Museum* 16: 523–646.

Raper, K. B. 1940. Pseudoplasmodium formation and organization in *Dictyostelium discoideum. J. Elisha Mitchell Sci. Soc.* 56: 241–282.

Raper, K. B. 1984. *The Dictyostelids.* Princeton University Press, Princeton, N.J.

Rose, M. R. 1991. *Evolutionary Biology of Aging.* Oxford University Press, New York.

Rose, S. M. 1970. *Regeneration.* Appleton-Century-Crofts, New York.

Sapp, J. 1994. *Evolution by Association.* Oxford University Press, New York.

Schmalhausen, I. I. 1949. *Factors of Evolution.* Blakiston, Philadelphia.

Schmitt, R., S. Fabry, and D. L. Kirk. 1992. Molecular origins of cellular differentiation in *Volvox* and its relatives. *Cytology* 139: 189–265.

Segel, L. A. 1984. *Modelling Dynamic Phenomena in Molecular and Cellular Biology.* Cambridge University Press, New York.

Shaffer, B. M. 1961. The cells founding aggregation centres in the slime mould *Polysphondylium violaceum. J. Exp. Biol.* 38: 833–849.

Shaffer, B. M. 1963. Inhibition by existing aggregations of founder differentiation in the cellular slime mould *Polysphondylium violaceum. Exp. Cell. Res.* 31: 432–435.

Simpson, G. G. 1953. The Baldwin effect. *Evolution* 7: 110–117.

Singh, K. S. 1992. *People of India: An Introduction.* Seagull Books, Calcutta.

Starr, R. C. 1970. Control of differentiation in *Volvox. Develop. Biol. Suppl.* 4: 59–100.

Steinberg, M. S. 1970. Does differential adhesion govern self-assembly processes in histogenesis? Equilibrium configurations and the emergence of a hierarchy among populations of embryonic cells. *J. Exp. Zool.* 173: 395–434.

Stern, D. L., and W. A. Foster. 1996. The evolution of sociality in aphids: A clone's eye view. In J. C. Choe and B. J. Crespi, eds., *Social Competition in Insects and Arachnids*, vol. 2, Evolution of Sociality. Princeton University Press, Princeton, N.J.

Tardent, P. 1963. Regeneration in the hydrozoa. *Biol. Rev.* 38:293–333.

Teissier, G. 1931. Étude éxperimentale du développement de quelques hydraires. *Ann. Sci. Nat. Zool.*, 10th series, 14: 5–60.

Thompson, D'Arcy W. 1942. *On Growth and Form.* Cambridge University Press, New York.

Thorsness, P. E., and T. D. Fox. 1990. Escape of DNA from mitochondria to the nucleus in *Saccharomyces cerevisiae*. *Nature* 346: 376–379.

Turing, A. M. 1952. The chemical basis of morphogenesis. *Phil. Trans. Roy. Soc. London* B 237: 37–72.

Viamontes, G. I., L. J. Fochtmann, and D. L. Kirk. 1979. Morphogenesis in *Volvox*: Analysis of critical variables. *Cell* 17: 537–550.

Waddington, C. H. 1957. *The Strategy of the Genes.* George Allen and Unwin, London.

Wheeler, W. M. 1911. The ant colony as an organism. *J. Morph.* 22: 307–325.

Williams, J. G. 1991. Regulation of cellular differentiation during *Dictyostelium* morphogenesis. *Current Opinion in Genet. & Develop.* 1: 358–362.

Wilson, E. O. 1990. *Success and Dominance in Ecosystems: The Case for Social Insects.* Ecology Institute, Oldendorf/Luhe.

INDEX

INDEX

JOHN TYLER BONNER is George M. Moffett Professor of Biology Emeritus at Princeton University. His books, *Life Cycles: Reflections of an Evolutionary Biologist, Cells and Societies, The Evolution of Complexity by Means of Natural Selection; The Evolution of Culture in Animals,* and *Size and Cycle: An Essay on the Structure of Biology,* are all available from Princeton University Press.